FOOD AND BEVERAGE CONSUMPTION AND HEALTH

# POMEGRANATE

## BOTANY, POSTHARVEST TREATMENT, BIOCHEMICAL COMPOSITION AND HEALTH EFFECTS

# Food and Beverage Consumption and Health

Additional books in this series can be found on Nova's website under the Series tab.

Additional e-books in this series can be found on Nova's website under the e-book tab.

FOOD AND BEVERAGE CONSUMPTION AND HEALTH

# POMEGRANATE

# BOTANY, POSTHARVEST TREATMENT, BIOCHEMICAL COMPOSITION AND HEALTH EFFECTS

ROBERT E. SMITH, PH.D.

nova publishers
*New York*

For permission to use material from this book please contact us:
Telephone 631-231-7269; Fax 631-231-8175
Web Site: http://www.novapublishers.com



**Library of Congress Cataloging-in-Publication Data**

ISBN: 978-1-63321-648-8

Library of Congress Control Number: 2014945911

*Published by Nova Science Publishers, Inc. † New York*

*This book is dedicated to my grandchildren*

# CONTENTS

# PREFACE

Pomegranate fruits along with their seeds, peels, flowers and juice have been consumed for thousands of years as foods and medicines. They contain dietary fiber, antioxidants, healthy unsaturated fats, minerals and other nutrients. They are grown around the world, but prefer a relatively warm, dry Mediterranean climate. The taste and aroma of the fruits and other parts of pomegranates depends on the age and maturity of the fruit, how it is processed after being harvested and how it is stored. They may be useful in helping to prevent and/or treat illnesses and conditions such as smoldering inflammation, obesity (metabolic syndrome), hyperglycemia, hypertension (high blood pressure), diabetes, stroke, heart diseases, several types of cancer and infectious diseases. However, unlike prescription drugs that have been through rigorously controlled clinical studies, pomegranates and products made from them are not prescribed or covered by health insurance. So, it is important to refrain from making false claims about the health effects of pomegranates. This is better understood if one has a sufficient background.

So, the book starts with an introductory chapter that provides some background about pomegranates and also describes the difference between proving something and getting a product approved by the US FDA or other countries' regulatory agencies. That is, a prescription drug must be manufactured in accordance with current good laboratory manufacturing practices (cGMP), tested for possible toxicity using good laboratory practices (GLP) and evaluated further using good clinical laboratory practices (GCLP). A very small percentage of all investigational new drugs (INDs) successfully navigate through the process and become approved drugs. Throughout the process, all procedures need to be well documented in writing. This is an essential part of western medicine. The spoken word is almost useless, in contrast to pomegranates and other foods which are selected by traditional methods. People often choose the foods juices and dietary supplements that they consume based on what they see and are told by parents, older siblings and friends. Verbal anecdotes, undocumented experiences and commercial advertisements are far more important influences on what people decide what to eat and drink than anything that is written in the scientific literature. So, all that one can say is that pomegranates and products made from them may help prevent and treat (but not cure) several diseases. However, this is best done as part of physically active, low-stress behaviors and a proper well-balanced diet. So, the importance of these factors in modern medicine is discussed in the introduction.

This is followed by chapters on the history and botany of pomegranates, production, postharvest treatment, biochemical composition and potential health effects. Since this includes some biology and chemistry, there is also an Appendix that provides a chemistry background for biologists, a biology background for chemists and a medicinal biochemistry background for both. That is, there is a relatively thorough discussion of phenolic antioxidants, DNA, RNA, proteins, lipids and carbohydrates. The chemicals and nomenclature mentioned in the chapter on biochemical composition are also clarified. The illnesses and conditions that pomegranates may be able to help prevent or treat are also described as well

*Chapter 1*

# INTRODUCTION

The pomegranate (*Punica granatum*) is the name of a beloved plant and fruit that has been used since ancient times [1]. The name comes from the Latin ''malum granatum", which means ''grainy apple". The name of the genus comes from the Roman's mistaken belief that it originated in Phoenicia [1]. It has been the subject of several reviews [1-10]. One of the older reviews [2] said that the enthusiasm for its health benefits may be only partly justified and is mostly due to its antioxidant properties. It also said that recent studies were beginning to suggest a possible synergism between the hydrophilic and lipophilic portions of pomegranates [2]. At the time that it was written the concept of synergism was controversial. Moreover, the importance of smoldering inflammation was not well understood. The following year a review article cited studies that showed the synergy that occurs when more than just the antioxidants in pomegranates are consumed [3]. Now synergism and smoldering inflammation are widely accepted in modern medicine, which is becoming a fusion of traditional and western medicine [11]. That is, traditional medicine is based on systems thinking, while western medicine is based on reductionist thinking.

One of the biggest differences in these two ways of thinking is this issue of synergism, in which the whole is greater than the sum of its parts. This is in contrast to reductionist thinking, in which the whole is equal to the sum of its parts. Until recently, very few peer-reviewed scientific journals accepted articles that described the biochemical activities or health effects of a mixture of compounds, a whole food or a juice. Such studies were often considered to be "dirty" and difficult to interpret. Fortunately, this trend is disappearing in modern medicinal chemistry [11]. "Before the 21$^{st}$ century, many doctors, biologists and chemists thought that reductionist and systems thinking were incompatible, but that is changing. Students learned reductionist thinking in elementary school, but are now learning that all things are connected to each other. That is, Science Fair projects in elementary school usually involve experiments in which one independent variable is altered at a time and one dependent variable is measured. In high school, college, graduate school and industry, adults realize that life is not that simple. In real life, many variables (or parameters) can change at the same time and cause multiple, nonlinear effects. Complex problems seldom have a single, simple solution. Many things should be done to prevent illnesses. This includes avoiding smoldering inflammation and the diseases that it can cause: heart disease, stroke, cancer, diabetes, autoimmune diseases and neurodegenerative diseases. This can be done in many people with a proper diet, physical activity, avoiding environmental toxins, regular physical examinations with a licensed physician and, when necessary, prescription drugs" [11].

So, let's look close at these two seemingly opposite ways of thinking and see how people are taking advantage of the best of each to optimize health care by doing many things, including consuming healthy foods, like pomegranates and products made from them. Still, we must remember that no one thing by itself can do the job and nothing – not even pomegranates - are of much use if a person chooses to consume tobacco products, subjects himself to excess UV light or engages in other behaviors that can cause many deadly diseases.

## Modern Medicine vs Traditional Medicine

Modern medicine is using the best of traditional and western thinking [11]. Each one has its own way of making decisions about health care. In traditional medicine, personal experience and verbal reports from trusted people are very influential. We make the decision whether or not to eat or drink something or take a dietary supplement based on this. Sometimes, we self-medicate. On the other hand, physicians and pharmacists decide what prescription drugs we should take. In traditional medicine, information was passed on to others verbally. Personal anecdotes were (and still are) thought to be important. In western medicine, only double-blind clinical trials are valid, and everything must be put in writing. Personal anecdotes about a food or drug being effective are considered to be unreliable. So, the FDA and regulatory agencies in other governments require pre-clinical and clinical studies that must be done in accordance with good laboratory practices (GLP) and good clinical laboratory practices (GCLP) before a new chemical entity (NCE) or investigational new drug (IND) can become an approved prescription drug [11]. Before such trials can begin, the NCE or IND must be made in accordance with current good manufacturing practices (cGMP) [11]. Even though results from such studies would never constitute proof in a logic class, they are used to decide if an IND has been shown to be safe and effective. If it is, the IND can become a legally prescribed drug and specific health claims can be made about it. This is relatively new in human history.

On the other hand, traditional medicine and natural remedies are older than humanity itself. Porcupines, chimpanzees, moths, ants and even fruit flies self-medicate with natural remedies [12]. They all observe nature and obtain healthy foods and medicines from it, just like our ancestors did. Long before there were printing presses or books, people learned though verbal communication. So, pomegranates and other important foods and natural remedies became popular through personal contacts. Usually, that is still how we choose them. Young children learn what to eat from their parents and older siblings. Adults decide what foods and dietary supplements to take by talking to friends, relatives and even their personal physicians. There are also many popular books about foods, beverages and "natural" remedies that are filled with personal anecdotes and stories, but have little or no information about well-documented clinical trials.

Still, eating habits and lifestyles emerged in traditional societies that helped maintain balance and health [11]. "They were formed through trial and error and a long verbal tradition. Most African, Asian and Native American diets are (were) rich in fruits and vegetables and had no man-made (or woman-made) chemical additives. Ancient foods and spices, such as pomegranates, turmeric and cayenne pepper were found to be important in maintaining good health and preventing disease. Some of them were identified as having healing properties. Unlike western medicine, no attempt was made in traditional societies to find a single active ingredient. Instead, entire portions of a food or herb were consumed. Sometimes several herbal remedies were given at the same time. In its extreme, holistic medicine teaches that an extract of a food or herb that contains a single chemical compound can never be as good as the mixture of compounds that exists in the food or herbal remedy. So, some people think that it is better to test the medicinal properties and toxicities of a whole food or herb, and not just pure compounds, since whole new properties can emerge with mixtures" [11].

On the other hand, we make decisions about which prescription drugs to take based on recommendations or orders from physicians, who must have a license to practice medicine. Unlike the food that we eat, prescription drugs must undergo a strict set of clinical studies that follow good laboratory practices (GLP) using drugs that are made under current good manufacturing practices (cGMP) [11].

## Safety and Toxicokinetics

The first step in developing a new drug is to give many different doses, including very large doses of the IND to test animals (usually mice or rats) to determine its toxicity and the lethal dose needed to kill 50% the animals, or $LD_{50}$. This is called a toxicokinetic study. Scientists also measure the concentration of the IND in the blood and various organs. They graph the concentration in the blood vs time and measure the area under the curve that is produced. This is called a toxicokinetic study. This establishes the half-life of the IND ($t_{1/2}$), time it takes to reach a maximum concentration in the blood ($t_{max}$) and the maximum concentration ($C_{max}$). The area under the curve (AUC) is a measure of the total amount of IND that is in the blood and selected organs. A simulated example is shown in Figure 1.

Scientists can estimate a safe dose that can be given to healthy people from the results of the toxicokinetic study in animals. So, the next step is what we call a phase I study, in which the IND is given to healthy people. One goal is to see if there are any adverse side effects and to do a pharmacokinetic study. The difference between it and the toxicokinetic study is that very high doses are not given, so there are no fatalities and hopefully, not even any harmful side effects. Also, one looks for a health effect instead of toxicity. Still, such studies are done in a facility, such as a hospital, so people can get immediate medical care should the need arise.

## Clinical Trials and Safety of Prescription Drugs

If the IND proves to be safe for healthy people, a phase II study can be done in which low doses of the IND are given to people who have the disease or condition that the IND is designed to treat. If the IND proves to be safe and effective to them, many more people are tested in phase III and IV studies. If the IND continues to be proven safe and effective in these studies, it may be approved by the FDA or similar agencies in other governments.

However, these studies can cost hundreds of millions of dollars and there is no guarantee that the IND will be approved or that a competing pharmaceutical company will get their own IND approved first. So, INDs and approved drugs are patented. They can be expensive, but health insurance will pay for them in most cases if they are prescribed by a licensed physician and used for their intended purpose.

These clinical studies must follow strict guidelines. The spoken word is seldom used. Everything must be written. It is important to say what you do, do what you say, and document it all in writing. Much of the initial research is driven by hypotheses, unlike traditional medicine. That is, scientists start with an idea that can be tested. The goal of their research is to confirm or deny the hypothesis. If investigators follow the proper guidelines and accumulate sufficient evidence, they can legally claim that they "know" that their hypothesis is true.

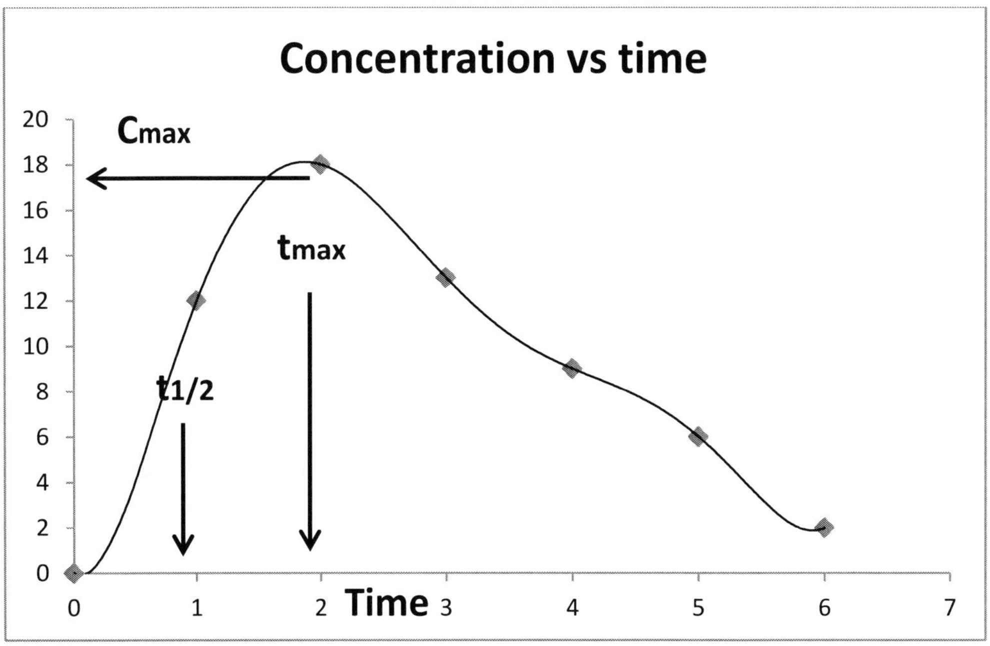


Figure 1. Simulated graph of concentration of an investigational new drug (IND) vs time after giving a dose of it to a test animal.

## SAFETY OF FOODS AND BEVERAGES

This is quite different from traditional medicine, in which one does not start with a pre-conceived idea. Also, foods and dietary supplements have many active ingredients, some of which may not be known. Foods and juices that have been consumed for thousands of years are generally regarded as safe, which is abbreviated as GRAS. Since printing devices and books have only been available for about 1000 years (starting in China), the decisions about what is edible and which herbs have medicinal properties were made without writing. So, the spoken word was used instead of the written word. Also, foods and juices are seldom ever tested on rodents in toxicokinetic studies. More importantly, natural products can't be patented, so there is seldom any incentive to do the expensive clinical studies that would be needed to "prove" that they have medicinal properties. Children just observe what their parents and older siblings consume and learn from that.

## EVIDENCE, PROOF AND KNOWLEDGE

So, in traditional medicine, one makes observations and draws conclusions from them. For example, in the seventeenth century Galileo used a telescope to see that there were four objects orbiting around the planet Jupiter. Even though it would be over 300 years before the laws of gravity would be established, he "knew" that the objects would continue to orbit Jupiter, thus "disproving" the idea that the Earth was the center of the universe and all objects in the sky orbit the Earth.

This is based on inductive reasoning. If the same thing is seen to happen many times, then it may be safe to assume that it will continue to happen. Indeed these four moons still revolve in orbits around Jupiter. Moreover, we now know that the Earth and other planets orbit the Sun. The law of gravity can be used to calculate the orbits of two objects, but the exact solution to the three body problem still eludes mathematicians and physicists. That is, we can't calculate the exact orbits of three bodies, such as the Earth, moon and Sun, let alone the 67 moons Jupiter. Still, NASA and other governments' space agencies do an excellent job of using the laws of gravity to calculate the amount and direction of thrust that is needed to launch a space craft into orbits around Jupiter and even Saturn. They "know" that they can make the necessary calculations because they work well enough to do what they want them to do.

However, in a logic class one learns that the only real proofs are based on deductive reasoning and syllogisms that are based on a premise. For example, one can start with the premise that all pomegranates produce a valuable, nutritious juice. You can hold a pomegranate in your hands and say that this is a pomegranate. The simple syllogism will tell you, "Therefore, this thing in my hands can produce a valuable, nutritious juice". The truth of the statement depends on the truth of the premise. So, even this form of knowledge is based on an assumption.

This points out the difference between "truth" and "knowledge" in a logic class compared to a science lab or a court room. Science can be used to help us survive and thrive, but it may not be able to establish anything with absolute certainty. In fact, one of the most fascinating things about the scientific process is that every answer to a question will usually raise more questions itself. For example, when Gregor Mendel discovered that some traits could be inherited, it immediately begged the question, "How are they inherited?" It took over 100 years to learn that they are inherited when we receive a pair of genes from our parents in the form of chromosomes that contain DNA. We also learned that evolution is not a matter of survival of the fittest, but survival of the most cooperative, despite the ideas that Darwin and many others suggested. In fact, there are no "favoured races" – there is only one race – the human race. The preservation of a species is not based on a struggle but on finding an ecological niche.

## CONTINUOUS IMPROVEMENT

In the 20th century, it was thought that all the information needed to produce an organism was in the DNA [11]. There was even a series of movies that suggested that scientists could clone dinosaur DNA in amber to produce exact replicas of extinct animals. Also, supposedly there was a one-way flow of information, or cause and effect. That is, genes make (or cause the production of) messenger RNA (mRNA), which make proteins, which make cells, which make organs, which make tissues, which makes humans as well as plants and animals. Because of the hierarchy, mRNA and proteins were not thought to cause the production of DNA. Diseases were thought to be caused by something going wrong with the living machine. To cure the disease, all that was needed was to identify the root cause of the problem in the machinery and eliminate it by giving a medicine with a single active ingredient. Indeed, many drugs approved by the FDA and other countries' medical regulatory agencies were thought to have a single molecular target. We were thought of as machines. Doctors were taught to avoid any emotional bond with the patient and to maintain an emotional distance, just as a car mechanic should avoid becoming emotional when repairing a car" [11].

So, in reductionist thinking we are like machines [11]. "Supposedly, to understand human biology and medicine we just need to understand all the individual parts of the human machine. Diseases are caused when something goes wrong, such as when a pathogen or toxin attacks one or more of those parts. If a doctor can identify the root cause of the disease, he or she can cure the disease. Doctors are supposed to focus on the disease and avoid any emotional connection with the patient. Certainly, this paradigm works well when treating a bacterial infection with an antibiotic. As a result, western medicine in the 20th century focused on finding root causes of diseases" [11].

At the end of the 20th century, reductionist thinking considered DNA to be the blueprint of life [11]. "It supposedly contained the complete code that is needed to make an organism. All the instructions for making a new human or for keeping him or her alive were said to be in the DNA. Information was said to flow from DNA to RNA and then to proteins, and not in reverse. A gene or piece of DNA codes for mRNA, which is translated into a protein. There was even a one gene one protein hypothesis. Each gene was thought to code for one protein. If one piece of the DNA (a gene) was defective, it would code for an mRNA molecule that would code for a defective protein. So, it should be theoretically possible to give the patient the proper gene and the disease would be cured" [11].

As a result, there was a huge international collaboration that solved the human genome. Scientists learned the sequence of all the deoxynucleotide bases (adenine, thymine, cytosine and guanine or A, T, C and G) in our 23 pairs of chromosomes [11].

However, once the sequence of bases was known, it raised many more questions. We learned that so-called identical twins can be quite different in their susceptibilities to diseases even though they are born with the same DNA [11]. "Even simple bacteria that have the same DNA can behave differently even when grown in the same environment (a cell culture). We also learned that there is a layer of control that lies above the DNA or genetics. It is called epigenetics, similar to the way that the word epicenter means the place on the Earth's surface that lies above the origin of the earthquake. We also learned that DNA is not a blueprint of life. The properties of DNA and the messages it conveys to a cell depend on the environment of the DNA and the cell or organism. Moreover, accessibility to the code can be affected by epigenetics that affect the accessibility of genes to transcription into mRNA. That is, we are coupled to the environment. We change our structure in response to environmental conditions, and we change our internal and external environments in response to our needs" [11].

Also, there is not a strict hierarchy of DNA causing mRNA to be made, which causes proteins to be made [11]. "There are also mobile genetic elements that can affect the phenotype, or the composite of an organism's observable traits, including health. Moreover, some pieces of DNA are transcribed into long pre-mRNA that contains insertion elements (introns) and exons. The introns are cut out and removed, while exons are spliced together. Exons can be mixed and matched in different ways and pre-mRNA modified further to make different mRNAs that are translated into different proteins. There are also other types of RNA (such as long non-coding RNA, lncRNA, and micro RNA, miRNA) that can affect the ability of a gene to be transcribed or an mRNA to be translated into a protein. Proteins that bind to RNA can control how DNA is transcribed. So, RNA molecules play central roles in gene expression and the generation of phenotypic complexity" [11].

Moreover, we continuously break down and re-make the most of the cells and parts of the organs in our bodies, unlike any machine [11]. "Our memories, structures, shapes and forms stay almost the same, even though the molecules have changed. Somehow, the most elderly people can remember things for over 100 years, even though none of the molecules in their bodies last for more than 10-20 years, and most last just a few days or weeks" [11].

## Autopoiesis and Emergent Properties

Another feature of life that illustrates systems thinking is the phenomenon of autopoiesis, or self-making [11]. "Machines do not make themselves. They are made by humans (or chimpanzees). On the other hand, almost all the cells in our bodies, and almost everything inside them, are continually being broken down and re-made, in a process called autopoiesis, or self-making. The turnover of blood cells in a man weighing 70 kg is close to 1 trillion cells per day, including 200 billion erythrocytes (red blood cells) and 70 billion leukocytes (white blood cells). This remarkable process of cell renewal is supported by a small population of bone marrow cells called hematopoietic stem cells. So, billions of human cells undergo programmed cell death, also known as apoptosis, every day. However, when cells don't die when they should, they can grow uncontrollably and form a tumor. So, many anticancer drugs stimulate apoptosis in cancer cells" [11].

Another important aspect of systems thinking is emergence [11]. "That is, properties appear, or emerge, when components of a system interact. For example, individual hydrogen (H) and oxygen (O) atoms have very different properties than they have when they are part of a $H_2$ or $O_2$ molecule. Also, completely new properties emerge when H and O become part of a water molecule. They emerge as the atomic orbitals interact and are transformed to make $H_2O$ molecules. Moreover, entirely new properties emerge when a hydrogen atom loses an electron to form a hydrogen ion, $H^+$, that is actually just a proton. This lowly proton is a crucial node in the network of life. Its concentration in living cells controls much of their lives and metabolism. Similarly, new properties emerge when ions and molecules interact to form living cells, which interact to form tissues, which interact in organisms, which interact with each other and with bacteria, fungi and viruses to form a human ecosystem or human body. Three corollaries to this are that there is seldom a single solution to a complex problem, everything is connected to everything else, and living organisms are not machines" [11].

Most people are aware of emergent properties, but have seldom thought of them in that way [11]. "For example, think about geometry. A single point contains a limited amount of information, but two points connected by a line segment have new, emergent properties. A traveler who is lost may know where he or she is located at a given point on a map. By itself, this is almost useless. He or she wants to know where the destination is and how to get to it. The destination is a second point, connected by a line segment to the first point. The properties of length and direction emerge as the two points interact at a higher level.

Next, let's consider chemistry. Although some people will say that it is based entirely on the principles of physics, we are only able to prove this with the hydrogen atom. The Schrödinger wave equation can only be solved for the hydrogen atom. Quantum mechanical calculations can only provide approximate solutions for all other atoms and molecules. These calculations require some assumptions. Thus, it is possible that the hydrogen in $H_2$ and the oxygen in $O_2$ have emergent properties that can't be explained entirely from the principles of physics. Better examples occur in biology. Many biologists consider the cell (and not atoms or molecules) to be the fundamental building block of life. The individual molecules, such as DNA, RNA, proteins, fats, sugars and small molecules (including water) have new properties when they are organized into living cells and organisms. New properties emerge as networks grow in size and complexity" [11].

## Systems Thinking vs Reductionist Thinking

So, systems thinking can be characterized by the following statement: "Genes are trapped inside huge colonies, locked inside highly intelligent beings, molded by the outside world, communicating with it by complex processes, through which, blindly, as if by magic, function emerges" [13]. "They are in you and me; we are the system that allows their code to be read; and their preservation is totally dependent on the joy we experience in reproducing ourselves. We are the ultimate rationale for their existence" [13]. In contrast, reductionist thinking can be characterized by an almost opposite statement, "Genes swarm in huge colonies, safe inside gigantic lumbering robots sealed off from the rest of the outside world ... They are in you and me; they created us, body and mind; and their preservation is the ultimate rationale for our existence" [14]. However, these are just metaphors, or philosophical visions. It is useful in guiding future research and in interpreting experimental results. Metaphors are useful in teaching and communicating, but are not scientific facts. It is impossible to think of any experiment that can "prove" that either of these contradictory metaphors is correct [13].

Perhaps a more useful metaphor for the genome is that it is like a CD or flash drive that contains a database that by itself, but does not code for anything [13]. Instead of a binary database, with ones and zeroes, like a CD, the genome is a quaternary database, with four monomers, or letters (A, T, G and C) instead of two numbers. Also, different parts of the genome database that are called genes can have very different meanings (or functions) depending on which cell or organism is accessing (or reading) the database. Genes can 'code' for very different proteins and types of RNA in different cells and different organisms. They can also code for very different proteins as cells grow or differentiate. Moreover, these proteins can have very different functions in different cellular environments. Genes are usually not 'selfish'. They usually do not determine the fate or behavior of the organism. In fact, the roles of many genes are affected by the environment of the cell, which is affected by the proteins (and other molecules and ions) that enter the cell from the environment [11, 13].

Another important feature of modern medicine is network theory [11]. That is, organs, cells, DNA, RNA, lipids, carbohydrates, proteins, small molecules and ions are all nodes that are connected in a network and we are all connected in the web of life [15]. This network is studied using math, science and engineering – all tools of western medicine [11]. However, the concept of everything being connected to everything else comes from traditional medicine [11].

## Fusion of Two SEEMINGLY Opposite Paradigms

So, even though reductionist and systems thinking may seem to be opposites and incompatible, they are both needed in modern medicine. Perhaps nothing epitomizes the fusion of traditional and western medicine more than "predictive, preventive, personalized and participatory (P4) medicine [16]. "It takes a holistic and quantitative, mathematical approach to practicing medicine. At the same time, systems medicine emphasizes prevention and individual participation in one's own health care. It recognizes the important human need for patients and care givers to be actively involved in preventing and curing diseases. At the same time, mathematics, the foundation of reductionist thinking, is used to quantify huge datasets from patients, while physicists, chemists, biologists and engineers develop the analytical tools needed to generate the data. All of this can be linked through the internet and used in mobile healthcare applications" [16].

At the same time, the North American Menopausal Society (NAMS), American College of Obstetricians and Gynecologists (ACOG), U.S. Preventative Services Task Force (USPSTF), National Institutes of Health (NIH), and US Food and Drug Administration (FDA) use evidence-based medicine in developing their guidelines for therapeutic practices [17]. Evidence-based medicine uses individual clinical expertise with the strongest available external clinical evidence that is obtained from systematic research. It has also been described as "the conscientious, explicit, and judicious use of current best evidence in making decisions about the care of individual patients" [18]. It involves reductionist thinking and systems thinking. That is, reductionist thinking is used in the basic science that is involved in establishing the proof of principle behind a drug's effects, improving the lead compound, studying pharmacokinetics and identifying diagnostic biomarkers. Systems thinking is needed when physicians "use their clinical experience to make an effective diagnosis and to thoughtfully identify and use compassionately individual patients' predicaments, rights, and preferences in making clinical decisions about their care" [18]. So, "evidence based medicine is not a "cookbook" approach, in which diagnoses and treatments are dictated from above. Instead, it is a bottom up approach, in which the best external evidence is used with individual clinical expertise and individual patients' needs" [18].

All of these factors are important in evaluating the health effects of pomegranates and products made from them. There are hundreds of bioactive compounds in pomegranates. They may be able to work synergistically with each other and possibly even some prescription drugs to help treat and prevent diseases. Now, in 2014, the concept of synergism is well established in pharmacy and toxicology. For example, isoniazid and rifampin work synergistically to treat tuberculosis [19]. Ribavirin and pegylated interferon-α work synergistically to treat hepatitis C infection [20]. Flubendazole and vinblastine work together to treat leukemia [21]. Pazopanib and paclitaxel synergistically treat anaplastic thyroid cancer [22].

## PROBLEMS IN SCIENTIFIC COMMUNICATION

Still, some problems remain – especially in communication. Unfortunately, every area of science develops its own vocabulary that can be confusing and even contradict normal vocabulary. For example, in organic and environmental chemistry there is an important toxin called benzene. It contains six carbons and six hydrogens that are arranged in a hexagon, as shown in Figure 2. The combination of solid and dashed lines between the letter C represent the hybrid bonds between the carbons and the solid lines between the C and H atoms represent the single bonds between them.

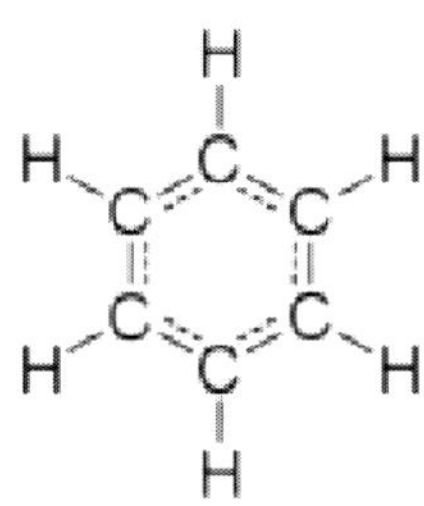


Figure 2. Structure of benzene, showing the carbons (C) and hydrogens (H), arranged in a hexagon.

It is more convenient to draw the structure as shown in Figure 3, in which the C and H are removed and the dashed lines are replaced by a circle, indicating that the electrons between the carbons are delocalized in what organic chemists call an aromatic compound.

Benzene can be converted to an acidic compound by adding an acidic functional group, such as an –OH, to make phenol, as shown in Figure 4.

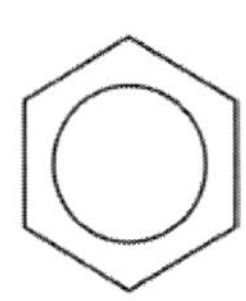

Figure 3. Structure of benzene, in which the carbons and hydrogens are not shown and the bonds between carbons are represented by a circle.

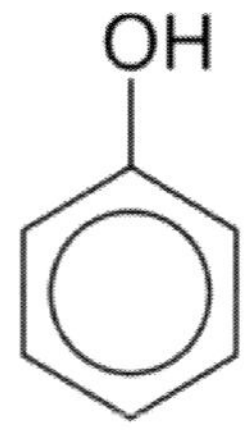


Figure 4. Structure of phenol.

More benzene rings and other functional groups can be added to make a variety of phenolic compounds, many of which are found in pomegranates and other fruits and vegetables.

Benzene can also be made acidic by adding a carboxylic acid, abbreviated as COOH. Sometimes it is often drawn as if there were a double bond between the carbon and one of the oxygens and a single bond between the carbon and the OH. Benzoic acid is the name of the compound. Its structure is shown in Figure 5.

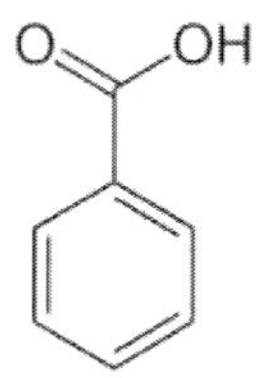


Figure 5. Structure of benzoic acid.

This illustrates one possible source of confusion. Instead of drawing the bonds between carbons as a circle, they are drawn as single and double bonds. Actually, the bonds between the carbons in the benzene ring have properties that are intermediates of hybrid of single and double bonds, so usually the benzene ring is drawn as it appears in Figure 4.

There is another source of confusion. Benzoic acid can be modified by adding an amine (such as or $NH_2$) to the benzene ring. If we used the same nomenclature that every other discipline uses and put the $NH_2$ on the carbon closest to the COOH, that would be the para position. That is, paralegals are nearly or almost real, legal lawyers. Paramedics are almost medical doctors. However, chemists use the italicized prefix *para* to indicate that the $NH_2$ (or any other functional group) is as far away from the COOH as possible. This is shown for *para* aminobenzoic acid (formerly the active ingredient in most sunscreen products) in Figure 6.

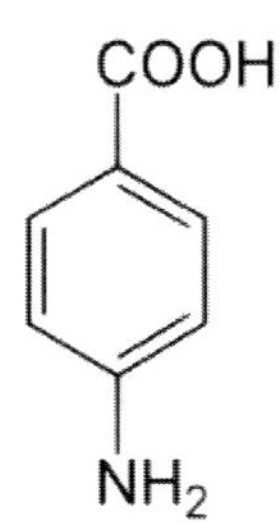


Figure 6. Structure of *para*-amino benzoic acid.

Also, the word acid is also frequently used and misused by food chemists. It is used properly when talking about benzoic acid or *para* aminobenzoic acid and fatty acids, such as palmitic acid, whose structure is shown if Figure 7.

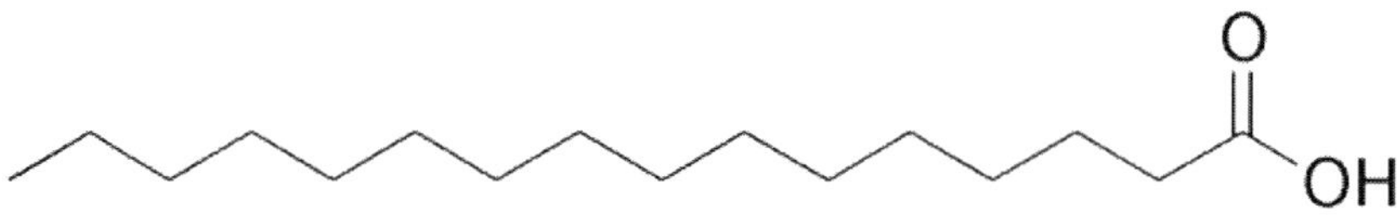


Figure 7. Structure of palmitic acid.

The zig-zag lines represent 15 carbons and 31 hydrogens arranged as follows: $CH_3CH_2CH_2CH_2CH_2CH_2CH_2CH_2CH_2CH_2CH_2CH_2CH_2$-. This is called an alkyl chain. Since there are no carbon-carbon double bonds (just carbon-carbon single bonds), palmitic acid is a saturated fatty acid. The carbons are saturated with all the hydrogens that they can hold and there are only single bonds between carbons. There are also many other fatty acids with different numbers of carbons. The alkyl chain on them can be represented by R. So, fatty acids have various numbers of carbons and hydrogens, with the carbons connected by single and sometimes double bonds. That is, some fatty acids are saturated, but some have carbon-carbon double bonds, so they are unsaturated. However, fatty acids (sometimes called free fatty acids) are seldom found in fresh food. Instead, there are triglycerides – especially in vegetable oils and in pomegranate seed oil. Triglycerides are esters. They can be made by reacting fatty acids with glycerol, as shown in Figure 8.

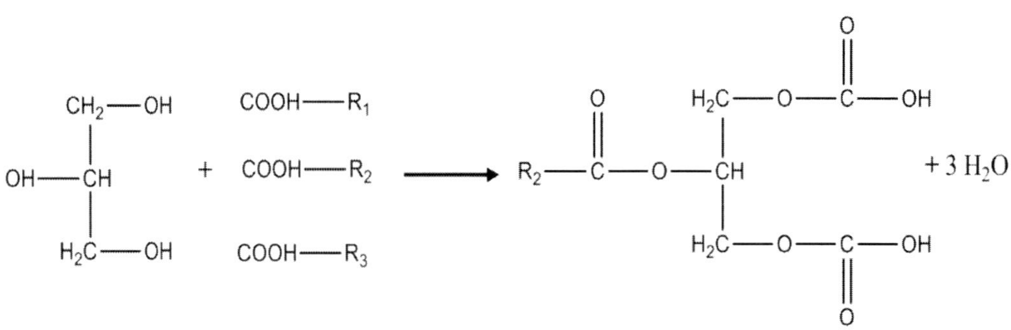


Figure 8. Reaction between glycerol and three fatty acids to make a triglyceride.

The fatty acids are converted to fatty acyls during the reaction. Fatty acyls and triglycerides have very different chemical, physical and medicinal properties than fatty acids. Unfortunately some of the scientific literature (including things written about pomegranate seed oil) doesn't make this distinction. In both the title and the main text they talk about the fatty acid content of vegetable oils and seed oils without mentioning triglycerides. In reality, only rancid oils have appreciable amounts of free fatty acids in them. They smell and taste bad and can cause indigestion.

The reason why such articles misuse the term is because of the way the authors analyze the oils. Gas chromatography (GC) is a relatively simple and inexpensive way of doing the analysis. However, as the name implies, GC is used to analyze volatile compounds and gases. Triglycerides are not volatile, so they must be converted to volatile derivatives before they can be analyzed. So, they are hydrolyzed in the opposite of the reaction shown in Figure 8. Fatty acids (and glycerol) are formed by the hydrolysis, but they are still not volatile. They are converted to volatile fatty acid methyl esters (FAMEs) by reacting the fatty acids with a form of activated methanol. Unfortunately, few analysts realize this, but it is simply human nature. In carpentry there is a saying, "Give a guy a hammer and pretty soon everything looks like a nail". A corollary to this is, "Give a guy a gas chromatograph and pretty soon everything starts to look like a fatty acid". However, if other methods, such as liquid chromatography with mass spectrometry (LC-MS) or nuclear magnetic resonance (NMR) are used, an analyst can directly observe the triglycerides and the different types of fatty acyls that are part of them [23, 24].

Fortunately, physicians read medical journals – not articles on food chemistry. Consumers read labels on foods. Medical journals and labels use terms like fats, saturated fats, *trans* fats, unsaturated fats and omega-3 fats. So, rest assured that if you have a problem with acid reflux and need to limit your consumption of fatty foods you can still eat fish, olive oil and pomegranate seed oil as well as take fish oil. They do not contain any acids.

GC and other methods of analysis will be discussed further in the Appendix, along with the different diseases and conditions that can be treated with prescription drugs and maybe even pomegranates and products made from them.

There is another common mistake in the scientific literature on food chemistry. Very few scientists do a good job of ensuring that they have extracted all the bioactive compounds in pomegranates (or other foods) before analyzing them. This is partly because very few journals will publish details on how extractions are done. This may be partly due to the misconception that bioactive compounds have similar properties when they are in foods compared to when they are separated from foods. That is, to do an analysis, scientists try to have pure standards available to calibrate their analytical instruments. As part of validating an analytical method, they will add some pure standards to the foods being analyzed, extract them and show that 100% of the standards can be solubilized in the extraction process. Many different solvents or mixtures of solvents can successfully solubilize all of the standards by simply shaking the mixtures of solvents and food samples at room temperature and pressure. However, this says nothing about whether they can penetrate cell walls in plants and completely solubilize the bioactive compounds. An example of this is a recent study [23] on the Asian fruit called noni, in which a very rigorous extraction method was able to solubilize much more material than previously reported by other workers. In that study [23], dry methanol at 100 $^{o}$C and 10 MPascal pressure (almost 100 atmospheres) was able to solubilize much more material. So, many of the previously reported values of concentrations of bioactive compounds in pomegranates and other foods could be severely underestimated. Moreover, many pomegranate products that are prepared by extraction could be extracted more thoroughly and produce much more material by using pressurized solvents at high temperature. Since some bioactive compounds (called anthocyanins) are easily hydrolyzed by water, dry methanol or ethanol should be used [23].

## Conclusion

So, let's look at the evidence about pomegranates and products made from them. It can be done in the same way that new prescription drugs are evaluated. First, the fruits must be made, so Chapters 2 and 3 will discuss the botany of pomegranates and production of thc fruit, juice and seed oil. This is analogous to how medicinal chemists learn to make a drug by chemical synthesis or purification from a natural source. Then, the biochemical composition of pomegranates and its health effects will be described in Chapter 4. This is analogous to identifying the active ingredients in a prescription drug. Finally, the potential health effects will be discussed in Chapter 5. Unlike FDA-approved prescription drugs, neither pomegranates nor any products made from them have been through all four stages of clinical trials needed to "prove" that they have therapeutic properties. However, as of 2013, 44 clinical trials were registered with the US National Institutes of Health (NIH) to study the effects of pomegranate juice or extracts on prostate cancer, benign prostate hyperplasia (BPH), diabetes, lymphoma, rhinovirus infection, the common cold, oxidative stress that occurs during diabetic hemodialysis, atherosclerosis, coronary artery disease, infant brain injury, hemodialysis for kidney disease, male infertility, aging, memory, complications during pregnancy, osteoporosis and erectile dysfunction. So, pomegranate juice and extracts are a little like new chemical entities (NCEs) in that they are in development and are still undergoing clinical trials to determine their safety and efficacy. None are available by prescription. However, in contrast to NCEs, the fruit, juice and products made from pomegranates can be purchased without a prescription. So, pomegranate juice and extracts are part of the fusion of traditional and western medicine. Readers can discuss with their physicians whatever applies to them so they can make the best decisions for their personalized health care. As our analytical techniques continue to improve, the effects of our diets on our genome, epigenome, transcriptome and metabolome will be measured. Results will be stored in patients' data clouds and decisions can be made to help predict and prevent diseases.

## References

[1] Holland, D. Hatib, K., Bar-Ya'akov. I. *Hort. Rev.* 2009, 35, 127-191.
[2] Lansky, E. P., Newman, R. A. *J. Ethnopharmacol.* 2007, 109, 177-206.
[3] Bell, C., Hawthorne, S. *J. Pharm. Pharmacog.* 2008, 60, 139-144.
[4] Viuda-Martos, M., Fernández-López, J., Pérez-Álvarez, J.A. *Comp. Rev. Food Sci Food Safety* 2010, 9, 635-654.
[5] Colombo, E., Sangiovanni, E., Dell'Agli, M. *EBC Alt. Med.* 2013, Article ID 247145, 11 pp.
[6] Da Silva, J. A. T. et al. *Sci. Hort.* 2013, 160, 85-107.
[7] Adhami, V.M., Khan, N., Mukhtar, H. *Nutr. Cancer* 2009, 61, 811-815.
[8] Nazamuddin, W. A. et al. *Int. J. Cur. Res. Rev.* 2013, 5, 16-22.
[9] Ismail, T., Sestili, P., Akhtar, S. *J. Ethnopharmacol.* 2012, 143, 397-405.
[10] Middha, S. K., Usha, T., Pande, V. *EBC Alt. Med.* 2013, Article ID 656172, 10 pages.

[11] Smith, R. E. *Medicinal Chemistry – Fusion of Traditional and Western Medicine*. Bentham Sciences, Sharjah, UAE, 2013, 687 pp.
[12] De Roode, J. C., Lefèvre, T., Hunter, M.D. *Sci.* 2013, 340, 150-151.
[13] Noble D. *The Music of Life: Biology Beyond the Genome*, Oxford University Press, Oxford, 2006.
[14] Dawkins R. The Selfish Gene, Oxford Press, Oxford, pp. 39, 214, 215 1976.
[15] Capra, F. *The Web of Life*, Doubleday, New York 1996.
[16] Flores, M. et al. Personalized Med 2013; 10: 565-576. P4 med
[17] Cirigliano, M. *J. Wom. Health* 2007, 16, 600-6631.
[18] Sackett, D. L. et al. *Brit. Med. J.* 1996, 312, 71-72.
[19] Cobelens, F. G. J. *Sci. Transl. Med.* 2013, 5(180fs), 1-3.
[20] Fried, M. W. et al. *Hepatol.* 2013, 58, 1918-1929.
[21] Spagnuolo, P.A. et al. *Blood* 2010, 115, 4824-4833.
[22] Isham, C. R. et al. Sci. Trans. Med. 2013, 5, 166ra3.
[23] Smith, R. E. et al. Noni composition and health benefits, in *Fruit Juices: Types, Nutritional Composition and Health Benefits*, Nova Sciences, Hauppauge, NY, 2014.
[24] Kaufman, M., Wiesman, Z. *J. Agr. Food Chem.* 2007, 55, 10405-10413.

*Chapter 2*

# HISTORY AND BOTANY

## HISTORY AND DISTRIBUTION

The history, botany and horticulture of pomegranates have been reviewed [1, 2]. The following is a summary of some of the key points. Pomegranates were first domesticated in the Transcaucasia-Caspian region and northern Turkey in the Neolithic era [1]. "Then they were cultivated in ancient Egypt, Greece, the Italian peninsula, Persia and what is now Iraq. Subsequently, they spread into what is now Turkmenistan, Afghanistan, India, China, North Africa and the Mediterranean region of Europe. Pomegranates were used in various ancient societies. The raw fruit, juice, peels and flowers were used as medicines, as described by Hippocrates, Pliny, Soranus and Dioscorides, and reviewed recently" [1]. Carbonized fragments of pomegranate peels dating from early Bronze Age have been discovered in Jericho and Arad, Israel; Nimrod, Lebanon; Egypt, and Armenia [2]. "Pomegranates were introduced by ancient societies in the Mediterranean region to the parts of Asia, North Africa and Europe. They also appeared on the Indian peninsula and Persia about the first century CE and were reported growing in Indonesia in 1416. The Greeks and the subsequent empires distributed pomegranates throughout Europe. Spanish explorers brought pomegranates to the Americas. Spanish Jesuit missionaries introduced pomegranates into Mexico and California in the 1700s. Since pomegranate trees can adjust to many different climatic conditions, they are found in the wild from Eurasia to the Himalaya Mountains" [2].

A Mediterranean-like climate is the best for growing pomegranates [2]. "This includes much exposure to sunlight, mild winters with temperatures not lower than 12 $^{o}$C, and dry hot summers without rain during the last stages of the fruit development. Under such conditions, the fruit will grow well and have optimal color and sugar content without the danger of splitting prematurely. However, when over-ripe they will spilt, as nature's way of helping to spread the seeds" [2].

"Commercial orchards of pomegranates can be found in Australia, the Mediterranean basin (North Africa, Egypt, Israel, Syria, Lebanon, Turkey, Greece, Cyprus, Italy, France, Spain, Portugal), Asia (Iran, Iraq, India, China, Afghanistan, Bangladesh, Myanmar, Vietnam, Thailand; and in the former Soviet republics of Kazakhstan, Turkmenistan, Tajikistan, Kirgizstan, Armenia, and Georgia), and the Americas (USA, Chile, Argentina and Brazil)" [2].

Mediterranean countries are the main centers for commercial cultivation, followed by Asian countries and countries of the former USSR [1]. "Cultivation is most efficient in Spain (18.5 t/ha), followed by the USA (18.3 t/ha). Iran exports the most (60,000 t/year), followed by India (35,176 t). Even though Spain has relatively little area (2000 ha) for cultivation, its share of exports is 37.8% of total production (37,000 t) followed by Israel (23.5%) and the USA (15.5%)" [1].

Through much of the 20$^{th}$ century, India imported pomegranate fruits from Afghanistan and Pakistan, but since the last decade of the 20th century, India has been exporting pomegranates to other countries [1]. It has ideal climatic conditions for growing high quality fruits throughout the year [1]. Throughout the world, growing techniques have improved while efficient industrial methods have been developed to separate the arils from the fruits [2]. Still, outcrossing occurs, so seedlings are not "true to type", resulting in plant-to-plant variations [1]. "Morphological changes can happen during domestication. As a result, cultivated forms often can be distinguished between their non-domesticated progenitors. However, pomegranate selections are made based on flower, rind and aril color, fruit size, sugar and acid contents, resistance to biotic and abiotic stresses, yield, post-harvest quality and the hardiness of the seeds" [1].

Various cultivars, landraces and varieties can be found in much of the world [2]. Landraces are cultivars that have been introduced into a different region. They adapt to their new climate and soil, so they often acquire new characteristics. The origin of some cultivars can be found in their names [2]. "The cultivars 'Kaboul' or 'Kandahary' in India probably originated in the Afghan cities of Kabul and Kandahar. Similarly, there are 'Afghansky', 'Washingtonsky', 'Iran 29-3', and 'Kalifornijsky' cultivars in Turkmenistan. Also, the term for pomegranate in Chinese is ''An Shi Liu'', which means ''the fruit of Kabul''. Still, it is difficult to verify the origin or authenticity of some cultivars, but they do have some distinguishing characteristics" [2].

The fruits of different cultivars have different characteristics that can be affected by the preferences of the consumers [2]. "For example, in India, most people dislike acidic fruit, so nonacidic cultivars have been selectively developed. In Israel, most people who immigrated from western European countries prefer sweet-sour cultivars, such as 'Wonderful'. On the other hand, Israelis originating from Middle East usually prefer nonacidic cultivars with very soft seeds such as 'Malisi'" [2].

Some of the better-known cultivars in India ('Ganesh', 'Mridula', 'Bhagwa') share common characteristics such as sweet flavor, low-acidity, small to medium fruit size and thin rinds [2]. "Many Indian cultivars originated from active breeding programs. 'Ganesh' is perhaps the best-known of them. It is an evergreen cultivar with very soft seeds and small fruits. However, if the fruits are thinned out, they can grow to as much as 350 g. The arils are red. They are sweet and have low acidity. The highest yield of marketable fruits occurs in January, but fruits can also be obtained in October, March-April, May-June, June-August, or July-September. Its skin color is green to orange-yellow, depending on the season. The yield and juice content is good, but depends on whether the trees are grown under optimum agricultural conditions. The yellow-green color of 'Ganesh', its tendency to split, and its lower quality are reasons why it is not exported much. Still, it has been used extensively in India for breeding and crossing with other cultivars, such as 'Kabul' (large fruit, yellow red skin, sweet, hard seeds); 'Jyoti' and 'Bedana' (medium-size fruit, brownish skin, sweet, soft seeds); 'Nana' and 'Kabul Yellow'. On the other hand, the Indian cultivars, 'Mridula'

('Arkta') and 'Bhagwa' ('Kesar') are frequently exported, especially to Europe. These cultivars have an appealing red skin and aril as well as soft seeds. They have low acidity and are sweet with a relatively small size (200 to 300 g). The peel is relatively thin, which can be a weakness, because it makes the fruits susceptible to physical damage. 'Bhagwa' is more prone to physical damage than 'Kesar'. 'Mridula', 'Bhagwa' and 'Ganesh' are evergreen cultivars. They are mostly exported to Europe - usually in January and February" [2].

The enormous size of the Indian peninsula and its very divergent climatic zones cause cultivars to adapt to the different regions [2]. "There are several other cultivars in India. This includes 'Alandi', 'Muskat', 'Jalore', 'Jodhpur Red', 'Dholka' (large fruit, yellow red skin, sweet, hard seeds, evergreen), 'Bassein', 'Malta', 'Kandhari' (large fruit, deep red skin, pink-blood red arils, subacid, hard seeds), 'Guleshah', 'Molus', 'Sharin', 'Anar', 'Jylothi', 'Bedana', 'Bosco', 'Srinagar Special', 'Chawla', 'Nabha', and 'Achikdana'" [2].

In the Islamic Republic of Iran, there are about 760 genotypes, specimens, and cultivars in the Yazd pomegranate collection [2]. "Since they were brought from many different provinces, synonyms or obvious similarities in appearance exist. Five valuable commercially important cultivars are 'Malas-e-Saveh', 'Rabab-e-Neyriz', 'Malas-e-Yazdi', 'Sishe Kape-Ferdos', and 'Naderi-e-Budrood'. They ripen late, produce medium to large sized fruits with thick red peels and red arils. Other cultivars include 'Ardesstani Mahvalat', 'Bajestani Gonabad', 'Ghojagh Ghoni', 'Khazr Bardaskn', 'Malas Yazd', 'Galou Barik', 'Bajestan', 'Zagh', 'Shavar Daneh Ghermez', 'Sefid', 'Togh Gardan', and 'Esfahani Daneh Ghermez'. 'Alack' is an early Iranian cultivar that ripens in late August to early September and is often exported. Its season lasts until the middle of September. 'Alak Shirin' (sweet) and 'Alak Torsh' (sour) are both are red, small and have hard seeds. 'Maykhosh' can be exported later. It can be picked until the end of December" [2].

As might be expected by such a large and diverse country, Chinese cultivars also show much variability and sometimes unusual features, such as being grown from spurs, and having double flowers or white flowers [2]. "They vary from small to very large and can be sour or sweet. Some of them produce fruits very early, beginning in early August, while some are later, with their season ending in November. There are also some evergreen cultivars. Most Chinese cultivars are either selections from unknown origin or are grown from seedlings obtained from known cultivars" [2].

The '87-Qing 7' cultivar is an early-bearing, productive spur-type cultivar [2]. "It is commercially important and its parent 'Qingpitian' might be of special help for gene mapping finding functional genes. The 'Duanzhihong' cultivar is another spur-type cultivar from Xingcheng. It is a compact bush. It ripens at the end of August and produces fruits that can weigh over 340 g and have a bright red skin color. The arils are a pinkish red. The 'Dabaitian' cultivar is sweet and has a white skin with white flowers. The 'Heyinruanzi' cultivar is sweet-sour, with a green skin and red flowers. The 'Tongpi' cultivar is both sweet and sour, with a green skin and red flowers. The 'Bopi' cultivar is sweet and sour with a red skin and red flowers. The 'Teipitian' cultivar is popular. It has a very large fruit that can weigh over 1.5 kg. It is yellow-green and the arils are bright red. There are two cultivars from the Lintong area in China, 'Linxuan 8' and 'Lintong 14'. 'Linxuan 8' ripens in September and has soft seeds. 'Lintong 14' matures in October. The 'Taishan Dahong' is an early cultivar from the southern foothills of the Tai Mountain in Shaanxi. One of the common pomegranate cultivars from the Sichuan province is called 'Qingpiruanzi'. It matures in mid- to late August. It produces a relatively large fruit with soft seeds and pink arils. Other cultivars with commercial importance in the Longyang district of the Baoshan municipality are 'Baishuijing', 'Chuanshiilu', and 'Hongshuijing'. There are also cultivars from Hui Li County in the Sichuan province. They are 'Ping Di' and 'Jian Di'. Both have green skin and soft seeds. The 'Yushiliu 1', 'Yushiliu 2' and 'Yushiliu 3' cultivars have red skins and have adapted to sandy alkaline soils. In China cultivars were primarily selected because of their size, juice content, seed softness, and time of ripening. There are also Chinese cultivars with unusual numbers of petals (''double flower'') and colors that are ornamental. They range from red-pink to pink-white. Some are even fertile and produce edible fruit. Examples of tasty double-flowered cultivars are 'Honghuachongbai' and 'Baihuachongbai'. The double-flowered 'Mudanhua' is similar to peonys (*Paeonia suffruticosa*) and are noted for their long flowering season, from early May to late October" [2].

There are also several cultivars in Garrygala, Turkmenistan [2]. "There are cultivars that ripen early, taste good and have an excellent composition of biochemical. The 'Kzyl Anar' cultivar produces relatively small fruits (200 to 250 g), hard seeds, and red or dark-cherry arils. The 'Acik-Dona' cultivar produces a high yield of medium to large fruits with hard seeds and pale crimson to red arils. Some of the Turkmen cultivars were sent to Israel and the USA" [2].

There are also several cultivars in Turkey [2]. "This includes the 'Cekirdksiz', 'Ernar', 'Fellahyemez', 'Hatay', 'Hicaznar', 'Izmir 1', 'Izmir 1264', 'Izmir 1265', 'Janarnar', 'Katrbas', 'Lefan', 'Mayhos II', 'Mayhos IV', 'Silifke Asisi', and 'Yufka Kabuk'. The 'Hicaznar'cultivar is red and produces many fruits that are sweet and sour taste with hard seeds, similar to the 'Wonderful' cultivar. 'Lefan' comes from Hatay. It has a yellow skin, large arils, a sweet-sour flavor, and very hard seeds. The 'Janarnar' cultivar has red skin, red arils, sweet-sour flavor, and hard seeds. 'Izmir 26' is sweet. So, most of the Turkish cultivars are sweet-sour, and red" [2].

Israeli pomegranate varieties are quite diverse considering the small size of the country [2]. "There are more than 50 pomegranate accessions Israel. Their fruits can have very divergent appearances, growth habits, ripening times, tastes and seed softness. The outside skin color varies from deep purple to yellow-pink, or green. Eight are grown commercially. All the Israeli accessions were given identity numbers because synonyms and homonyms were likely. Based on comparisons in a single orchard in Newe Ya'ar, some cultivars are very different from each other but have the same name, such as 'Hershkovich', while others that are identical have different names, such as 'Wonderful', 'P.G.100-1' and 'P.G.101-2', all of which have commercial value, as do the following: 'P.G.116-17', 'P.G.128-29' ('Akko'), 'Shani-Yonay', 'Rosh Hapered', 'P.G.127-28' ('Black'), 'P.G.118-19' ('Hershkovich'), and 'Malisi' ('P.G.106-7')" [2]

Traditionally, 'Rosh Hapered' 'Malisi', and 'Red Lufani' ('Shara'bi') have been grown in Israel [2]. "The first two cultivars have pink arils and taste sweet and not sour. 'Rosh Hapered' produces large fruits and arils, hard seeds and a pink skin. It is used in the Jewish holidays because it ripens at the end of August and is exported. On the other hand, the 'Malisi' cultivar produces fruits with soft seeds. It has light pink arils and yellow-pink to green skin. It is grown only in small plots and is not exported. 'Red Lufani' is a synonym of 'Wonderful', as is the main cultivar, 'P.G.101-2', which imported from the USA about 100 years ago. It produces large fruits and ripens in the beginning of October. It is has a sweet-sour flavor as well as red arils and a pericarp (peel or skin) when fully ripe. The different landraces ripen at different times and have a variety of external colors, timing of skin color appearance during fruit development and different seed hardness. Among the 'Wonderful' landraces, 'Kamel' is considered to be the most colorful. Its peel is red and it develops much earlier than the regular 'Wonderful' landrace. It is very productive and produces high-quality fruits" [2].

As the export market and demand for early red cultivars increased, two more early red cultivars, 'Akko' and 'Shani-Yonay' emerged [2]. "They have soft seeds with a sweet/sourless taste and red skin. Their appearance and taste make them the leading cultivars for early export. The growth habit of the fruits and trees of the 'Akko' cultivar differ from those of 'Shani-Yonay', but they both produce fruits that weigh 300 to 400 g. There are also two other cultivars, 'P.G.116-17' and 'P.G.118-19' ('Hershkovich'), that ripen between the late 'Wonderful' and the earlier cultivars. 'P.G.116-17' is now the most exported cultivar. It produces a large fruit, large red arils, and a red rind. Another Israeli cultivar that has been grown commercially is the black 'P.G.127-28'. It has a deep purple-black skin and red, soft-seeded arils. It produces a small fruit that matures in November in the microenvironment of the Newe Ya'ar Research Center and is the latest ripening cultivar tested. Its unusual skin color, very late ripening time, and soft red arils make it an appealing cultivar" [2].

There are at least 40 Spanish cultivars that have been divided into three goups: sweet, sweet-sour and sour [2]. "Some of the popular commercial cultivars are 'Mollar de Elche' and its selections 'ME1' ('Mollar de Elche No. 1'), 'ME5', 'ME6', 'ME14', 'ME15', 'ME16', and 'ME17', 'Agria de albatera', 'Agria de Blanca', 'Agridulce de Ojos' ('ADO'), 'Albar de Bianca' ('BA'), 'Borde de Albatera' ('BA') and its selection 'BA1', 'Borde de Blanka' ('BB'), 'Casta del Reino de Ojos' ('CRO') and its selection 'CRO1', 'Mollar de Albatera' ('MA') and its selection 'MA4', 'Mollar de Orihuela' ('MO') and its selection 'MO6', 'Pinon Duro de Ojos' ('PDO'), 'Pinon Tierno Agridulce de Ojos' and its selections 'PTO1' ('Pinon Tierno de Ojos No. 1'), 'PTO2' and 'PTO7', 'San Felipe de Bianca' ('SFB'), and 'Valencian No. 1' ('VA1'). There are also some Spanish cultivars like 'Mollar de Elche', 'Borde de Albatera', 'Pinon Tierno de Ojos' and 'Casta del Reino de Ojos' that have several selections or clones. The different numbers of the cultivars indicate that there are different landraces within each cultivar" [2].

The most popular Spanish cultivar is the landrace 'Mollar de Elche' [2]. "It produces sweet fruit with soft seeds. The outer color is pinkish red and the arils are red. It ripens in October and November. 'Valencian' is another sweet cultivar. On the other hand, 'Agridulce de Ojos' and 'Pinon Tierno de Ojos' are sweet-sour, while 'Agria de Albatera', 'Agria de Blanca', 'Borde de Albatera' and 'Borde de Blanca' are sour. 'PTO2' and 'CRO1'

Produce fruits with a high juice content" [2]

In contrast, there are only a few cultivars in the USA [2]. "'Wonderful' is the most important. It originated in Florida, but was later discovered in Porterville, California, at about 1896. It is the most widely planted commercial cultivar in California. It produces large fruits with large red arils. It has a sweet-sour taste, has semi-hard seeds, and survives well when shipped to other places. The fruit has a very appealing red, glossy color. There are also several Israeli landraces of 'Wonderful', but it is not clear whether the American 'Wonderful' is genetically distinguishable from any of the Israeli 'Wonderful' landraces. However, the American 'Wonderful' fruits are much harder and less prone to mechanical extraction of arils than the Israeli landraces. These differences could reflect variations in growth conditions and/or environment. Moreover, 'Wonderful' is grown in western Europe and Chile" [2].

'Early Wonderful' is a sport of 'Wonderful' found in California [2]. "It ripens about two weeks before 'Wonderful' and it becomes red much earlier than the original cultivar. Also, the quality of the juice is inferior to that of the original 'Wonderful'" [2].

Another early commercial cultivar with red skin and aril color is 'Early Foothill' [2]. "It is much smaller than the original 'Wonderful' and its fruit quality is not as good. 'Granada', a 'Wonderful' sport, is a cultivar in the USA that ripens in mid-August. The quality of the fruit and juice is not as good as other cultivars and its commercial value is limited. 'Ruby Red' is a cultivar with similar size and ripening time as 'Wonderful' but has a shorter shelf life than 'Wonderful'. Other cultivars grown to a limited extent in the USA are 'Balegal', 'Cloud', 'Fleshman', 'Crab', 'Francis', 'Green Globe', 'Home', 'King', 'Phoenicia', 'Sweet', and 'Utah Sweet'. Several ornamental pomegranate cultivars, such as the 'California Sunset', are also sold in the USA. At least two originated in Japan, including the ''double-flower'' cultivars: 'Nochi Shibori' and 'Toyosho'" [2].

Several cultivars are also found in the former Soviet Republic of Georgia [2]. "They are 'Pirosmani', 'Gruzinskii No. 1', 'Gruzinskii No. 2', 'Vedzisur'i, 'Lyaliya', 'Tengo', 'Imeretis Sauketeso', 'Bukistsikhe', 'Khorsha', 'Zugdidi', 'Erketuli', 'Forma No. 1', 'Forma No. 15', 'Forma No. 70', 'Shirvani', 'Apsheronskii Krasnyi', 'Burachnyi', 'Rubin', 'Frantsis' 'Sulunar', 'Kyrmyz Kabukh', 'Shiranar', Shakhanar', and 'Gyuleisha Krasnaya'. The 'Apsheronskii Krasnyi', 'Burachnyi', 'Frantsis', 'Kyrmyz Kabukh', 'Lyaliya', 'Pirosmani', 'Rubin', 'Shirvani', and 'Verdzsuri' cultivars are resistant to splitting. The 'Sulunar', 'Pirosmani', 'Vedzisuri', and 'Imeretis Sauketeso' cultivars have the highest juice content" [2].

Tunisia also produces several cultivars, but most are consumed locally and are not exported [2]. "This includes the 'Gabsi' (the main cultivar, sweet); 'Tounsi' (sweet, late ripening); 'Zehri' (sweet, ripens end of August or beginning of September); 'Chefli' (sweet, poor skin color, big nice arils); 'Mezzi', 'Jebali', 'Garoussi' (sweet-sour, green skin); 'Garoussi'; 'Kalaii' (sweet, poor skin color, big nice arils); 'Zaghouani'; 'Andalousi' (sweet); and 'Bellahi' cultivars. There are also four cultivars in Egypt: 'Arabi', 'Manfaloty', 'Nab ElGamal', and 'Wardy'. 'Manfaloty' was more sensitive to salt stress while 'Nab ElGamal' lost less chlorophyll when irrigated with water containing an elevated concentration of salt. The 'Manfaloty' (or 'Manfaloot') cultivar produces large, juicy dark-red arils and ripens from the end of August to the beginning of September. The Egyptian 'Granada' is an early cultivar. It is not clear whether it is identical to the American 'Granada', although its early ripening season suggests that they are similar" [2].

"There are also six Sicilian or Italian cultivars ('Dente di Cavallo', 'Neirana', 'Profeta', 'Racalmuto', 'Ragana', and 'Selinunte'), 17 clones and cultivars in Morocco ('Gjeigi', 'Dwarf ever Green', 'Grenade Jaune', 'Gordo de Javita', 'Djeibali', and 'Onuk Hmam'), three in Iraq ('Ahmar' (red), 'Aswad' (black), and 'Halwa' (sweet), and one each in Saudi Arabia ('Mangulati') and Vietnam ('Vietnamese' – an evergreen)" [2].

## Botany and Genetics

The genus *Punica* has only two species, *P. granatum* and *P. protopunica* [1]. Based on the anatomy of the xylem, *P. protopunica* has been suggested to be the older of the two [2].

"*P. protopunica* is only found on the Socotra Islands in Yemen. However, *P. granatum* is grown in many different geographical regions including the Mediterranean basin, Central Asia, China, India, South Africa, Australia, South Africa and the Americas. The color of the pomegranate ovary is a stable characteristic that is retained when plants are grown from seeds" [1]. "It has been used to distinguish two subspecies of *P. granatum*, chlorocarpa (found mainly in Transcaucasia) and porphyrocarpa (found mainly in Central Asia). Today, pomegranates are cultivated throughout the world in subtropical and tropical areas in many variable climatic conditions, indicating its flexibility, adaptability, and wide range of genetic diversity" [1].

The acceptability of pomegranates by producers and consumers depends on the color of the peels (rinds), sugar content and acidity [3, 4]. Spanish researchers studied the relationship between the color of the peel during fruit development and maturation and the air temperatures. A correlation coefficient of 0.9 indicated that air temperature had a significant effect on the development of the color of the peels [4]. "Color was measured using color coordinates or parameters. The lightness coordinate ($L^*$) increased continuously from when it was young (second week of June) until it reached a maximum during the second week of September, when the air temperature was highest. After that, the lightness coordinate decreased until the pomegranates were harvested in late October. The color coordinate $a^*$ was negative, indicating a green color until the second week of September, when it turned positive, indicating a red color. This coincided with the maximum air temperature and $L^*$ coordinate. After that, $a^*$ increased (more intense red), while $L^*$ decreased. The maximum $a^*$ values were seen between the first and second week of October. The green color of the peel changed to red during this time. The yellow/blue coordinate ($b^*$) was a high positive value (indicating yellow) while the fruits developed and ripened. From the second week in September onward, the $b^*$ value decreased as the blue color replaced the yellow. In addition, the color of the peel was strongly correlated with air temperature, as all color coordinates increased. This included the saturation index ($C^*$, also known as chroma index) and hue angle ($h_{ab}^*$)" [4]. They are used to describe color in a way that agrees well with visual experience [5]. The chroma index and hue angle indicated when the color of the peels turned from green to red [4].

The changes that occur as pomegranates grow and mature were reviewed recently [6]. "The importance of properly timing the harvest was emphasized. Pomegranates are non-climacteric fruits, meaning that they ripen without a burst of ethylene production and increase in cell respiration. There are differences between cultivars. For example, flowers can be male, hermaphrodite or intermediate. They can also be solitary or grouped. There are 43-66% male flowers in the Israeli 'Wonderful' cultivar and 78-86% for the Turkish 'Hicaznar' cultivar. The rate at which the fruits grow can also vary. The 'Malas-e-Torsh-e-Saveh' cultivar produces fruits in which the average weight and volume increases rapidly until 45 days after the fruits first set and then continues more slowly until they are harvested. The growth of the 'Mule's Head' cultivar followed a sigmoid pattern, but the 'Wonderful' cultivar had linear growth with respect to time. In the hot valleys in Israel, fruits matured faster than on the cooler plains. The fruits ripened 5-8 months after the fruit firsts set. They change from flowering to maturity and then senescence. These changes are accompanied by structural, biochemical, physiological and elemental changes. The weight of the fruit of South African 'Bhagwa' cultivar increased from 107 g 54 days after the full bloom of the flowers to 322 g when harvested 139 days after full bloom. On the other hand, the fruits of the Australian

'Wonderful' cultivar had a nearly linear rate of growth. The fruits weighed about 675 g 14 weeks after the fruit first set. The length and diameter of the fruit increased while the ratio of length to diameter (shape index) decreased during development and maturation. This is because the fruits continue to grow even after the optimum time for harvesting, presumably due to cell expansion caused by the uptake of water and nutrients. Isreali pomegranates are smaller when the fruits ripen in early summer or during the winter than when they ripen in the late summer or autumn. The peels of the Iranian 'Malas Yazdi' cultivar contributed more to the total mass of the fruits early in the growing season, while the arils became dominant from the middle to end of the season with a mean dry weight of 35.03 and 22.33 g in arils and peel, respectively, during late parts of the season. On the other hand, the fruit arils were 50% of the fruit weight during most stages of fruit development for Israeli 'Mule's Head' and 'Wonderful' cultivars, while 57–66% of fruit weight was due to arils in the Spanish 'Mollar' cultivar. In the South African 'Bhagwa' cultivar, fruit arils were 50% of the weight of the fruits until 110 days after full bloom when they were semi-ripe and increased to 58% when fully ripe" [6].

Juice from the Australian 'Wonderful' cultivar was 37% of the total weight of the fruit [6]. "In another study, the percentage of juice from the 'Mule's Head' and 'Wonderful' cultivars ranged from 35-40% and 40-45%, respectively. The percentage of juice from the 'Wonderful' cultivar grown in Israel ranged from 18-40%, due to being grown in different climatic conditions. The pomegranates grown in the early summer had the highest juice content. The juice content of the South African 'Bhagwa' cultivar increased from 29% to 54% as it advanced from the immature to full-ripe stage. Omani cultivars produced 57-67% juice" [6].

The color of the fruit is also an important factor that affects the acceptability of pomegranates [6]. "However, often there is no correlation between the color of the peels and that of the arils. For example, the arils of the Spanish 'Mollar de Eche' cultivar change from white to pinkish-red while the peels change from green to greenish yellow, and finally to brownish yellow with reddish patches. Also, the color of the juice is influenced by pre- and postharvest conditions. Fruits grown on the coastal plains developed a more intense color than those grown in warmer valleys in Israel. The intensity of the red color of the arils was less when pomegranates were grown at higher temperatures. During the middle of the growing season, the color of the arils gradually changed from white to pink as the fruit ripened" [6].

The concentration of total soluble sugars (TSS) also increases during maturation [6]. Most articles and books use the term total soluble solids instead of total soluble sugars. The TSS value is obtained by measuring the refractive index of the juice or an aqueous extract of the fruits using a refractometer. It measures the amount (in degrees) that light is refracted when it passes through a sample. This detects sugars but not other compounds that might be extracted. It is measured in units of degrees Brix or $^{o}$Brix, which can be converted to percent TSS. It does not include triglycerides or other lipophilic compounds that can be extracted from seeds using hexane or other solvents. "The two main sugars in pomegranates are fructose and glucose. The TSS of the 'Bhagwa' cultivar increased about 150% between 54 days after full bloom until being harvested 165 days after full bloom. Similarly, the TSS of the Indian 'Ganesh' cultivar increased from 13% in 40-day old fruit to 16.3% on day 140. In contrast, the TSS only increased from 16.4 to 16.9% in the 'Taifi' cultivar and remained nearly constant throughout the ripening process in the Spanish clones 'ME5', 'ME17' and 'MO6'and sweet 'Mollar' cultivar" [6].

Titratable acidity is also important [6]. "The 'Wonderful' cultivar has a relatively high acidity. Usually, titratable acidity decreases as pomegranate fruits mature, but the rate of decline can vary between different cultivars. For example, the titratable acidity of the South African 'Ruby' and 'Bhagwa' cultivars from 0.39 to 0.31% and from 0.62 to 0.38% respectively from day 54 after first bloom to either day 139 ('Ruby') or day 165 ('Bhagwa'). As the acidity decreases, the pH of the juice increases. Depending on the cultivar and degree of maturity, the pH can be between 3.2 and 4.2. Also, as the total acidity decreases, the TSS increase" [6].

The concentrations and types of organic acids can also affect the perception of sweetness and sourness [6]. Citric acid is the most abundant in almost all cultivars. The concentrations of other acids vary between different cultivars. These include tartaric, malic and oxalic acids. In the 'Ruby' cultivar tartaric acid is the most abundant. Another minor but very important acid is vitamin C (ascorbic acid). Its concentration tends to decrease during maturation. The ascorbic acid concentration in the 'Rabbab-e-Fars' cultivar decreased from 25.84 mg/100 g in 20 day-old fruit to 9.78 mg/100 g in 140 day-old fruit" [6].

Astringency, caused primarily by the presence of phenolic compounds, is another important factor in the acceptability of pomegranates and the juice made from them [6]. "As fruits mature, the astringency and phenolic content decrease. Decreases of about 50% have been reported. When fully mature, the 'Rabbab-e-Fars' cultivar had 786 mg/100 g of total phenolics" [6].

One group of phenolic compounds, anthocyanins, gives pomegranates their red color [6]. The same anthocyanins are seen in most cultivars, but their relative concentrations can vary. This includes the 3-glucosides and 3,5-diglucosides of pelargonidin, cyanidin and delphinidin. The total amount and concentration of anthocyanins tends to increase during maturation. Spanish pomegranates have been found to have a five to ten-fold increase in anthocyanins as they grow from an immature to mature state, with delphinidin 3,,5-diglucoside being the most abundant" [6].

The concentrations of metals also changes during maturation [6]. "Potassium ($K^+$) is the most abundant. Its concentration is higher in the arils of unripe fruits, but is lower in the juice obtained from ripe fruits. The concentrations of phosphorus (P), sodium and calcium also increased in arils, while magnesium ($Mg^{2+}$), sodium and calcium in juice decreased with advancing maturity. In the peels, the concentrations of P, $Mg^{2+}$, manganese, zinc and copper were highest in the unripe and immature stages in the 'Hicaznar' cultivar" [6]. The authors said that the mature and unripe fruits had an equal amount of $K^+$ [6], but it was actually the concentration that did not change. The amount increased because the size of the fruits increases during maturation. Amount is equal to the concentration times the weight of the fruit or volume of the juice.

There are also important physiological changes in pomegranates as they mature [6]. As mentioned previously, they are non-climacteric. The respiration rate decreases as the fruits develop and ethylene gas is not produced [6].

When the review article was published in 2013, the authors said that there was no reliable maturity index [6]. In 2014, though, other authors described a minimum maturity index based on a fresh weights of the fruit and arils, along with a decrease in hydrolyzable tannins, based on studies of two Israeli cultivars, 'Acco' and 'Shani-Yonay' [7]. The weight and concentration of hydrolyzable tannins correlated well with an improvement in flavor and a decrease in undesirable astringency [7]. They concluded that a "weight of >0.23 g for the arils would make a simple and reliable index of maturity for timing the commercial harvesting of these two early-season "sweet" pomegranate varieties" [7]. As one might imagine, this can be influenced by the genetics of the pomegranates.

The genetic diversity of pomegranates is indicated by the existence of over 500 varieties [1]. Most varieties have 8 pairs of chromosomes, but the 'Double Flower' variety has nine [8]. "Also, some have up to five B chromosomes (also called accessory or supernumerary chromosomes). They are not essential for life. Pomegranates can be bred to give them desirable properties, such as having soft seeds, which is controlled by several genes. However, the genes coding for hard seeds are dominant, as are those coding for red and pink arils. Also, the acidity of the fruit is primarily controlled by one gene, with sourness or acidity being dominant over sweetness and low acidity. A gene has also been found that controls the colors of the petiole base, leaf margin, flower bud and fruit rind, with red ('Ganesh') being dominant over yellow ('Kabul Yellow')" [8].

Pomegranates can be bred to give them desirable properties, such as having soft seeds [8]. "This trait is controlled by several genes. However, the genes coding for hard seeds are dominant, as are those coding for red and pink arils. Also, the acidity of the fruit is controlled mostly by one gene, with sourness or acidity being dominant over sweetness and low acidity. A gene has also been found that controls the colors of the petiole base, leaf margin, flower bud and fruit rind, with red ('Ganesh') being dominant over yellow ('Kabul Yellow'). There is also the 'Double Flower'. It is an ornamental variety in which the numerous stamens are modified into petals, resulting in attractive flowers, resembling a rose" [8].

Pomegranates are diploid. Most varieties have 8 pairs of chromosomes and many have accessory B chromosomes [1]. In metaphase, the lengths of the chromosomes range from about 1.2 to 2.2 μm, with some variation between 11 Egyptian cultivars [9]. During the diplotene stage of meiosis, homologous non-sister chromosomes can exchange genetic material as they form DNA crossover points called chiasma [10]. They can be used along with accessory B chromosomes to measure genetic diversity in pomegranates [11]. In a study of 21 Iranian cultivars, it was found that not only the expected bivalent (diploid) chromosomes were seen during metaphase I of meiosis, but also some tetravalent ones [11]. This should not be entirely unexpected, since even a small percentage of human liver cells can be tetraploid. In pomegranates, a positive correlation was found between the amount of annual rainfall and number of ring bivalent chromosomes [11]. "Altitude had a negative correlation with the number of univalent or single chromosomes. Longitude had a positive correlation with the number of ring bivalents and a negative correlation with the number of univalents and intercalary chiasmata. The presence of B chromosomes increased the number of chiasma in some varieties, but decreased it in others. The cultivars also differed in the types and frequencies of chromosomal abnormalities, which may be useful in selecting cultivars and making hybrids. Sometimes their chromosomes can be sticky, thus producing diploid ($2n$) gametes and giant grains of pollen. Overall, statistical analysis of meiotic properties enabled investigators to group different varieties of pomegranates based on cytogenetic diversity" [11].

Still, it is important to begin looking at the DNA in the chromosomes more closely to find better genetic markers. Pomegranate DNA, like human DNA, can be analyzed to see which parts of it contain genes that are expressed as proteins. That is, protein-coding genes are translated into mRNA, which is translated into proteins. To determine which parts of the genome contain protein-coding genes, complementary DNA (cDNA) is made from the mRNA in a cell. The cDNA is used to clone or make many copies of DNA in the initial stages of sequencing a genome. Short pieces of the cDNA called expressed sequence tags or ESTs can be used to locate genes on chromosomes. ESTs are regions of chromosomes that are transcribed, or expressed as mRNA and proteins. So, there is a large public database (NCBI dbEST) that contains over 45 million ESTs [12], including 787 from pomegranates. [13]. These ESTs were used together with simple sequence repeats (SSRs) to identify genetic markers [13]. "The ESTs were assembled by compotrer analysis using CAP 3, a DNA seuqnce assembly program. It found 415 genes. Fifty-nine SSRs were obtained from these genes and used with ESTs to form primer pairs for cloning and analyzing DNA. This produced 18 genetic markers, 15 of which were polymorphic among 42 pomegranate accessions. The heterozygosity and polymorphic information content ranged from 0.119 to 0.619 and from 0.091 to 0.656, respectively. The analysis was able to identify distinct markers for pomegranates from east-central China" [13].

SSRs were also used to analyze 738 pomegranate accessions from 23 provinces in Iran [14]. "Forty-three alleles were detected. The heterozygosity and polymorphic information content were 0.521 and 0.458, respectively. Statistical analysis grouped the pomegranates into eight main groups, with geographical location being important. Finally, enough genetic variety was seen to encourage further efforts to breed new varieties and manage germplasms" [14].

Genetic markers can also be obtained by random amplified polymorphic DNA analysis, or RAPD [15]. "This method uses random DNA sequences containing 10 base pairs to reveal polymorphisms in a genome. Probes and filters do not have to be prepared and DNA does not have to be sequences, making RAPD relatively easy. So, Turkish pomegranates were analyzed using 47 RAPD primers to identify nine polymorphic patterns" [15].

The genetic varieties of Iranian pomegranates were analyzed by RAPD, SSR and inter-simple sequence repeat (ISSR) genetic markers [16]. "By using all three types of markers, more genetic diversity could be seen. Homonymous, synonymous and/or mislabeled genotypes were identified. Pomegranates were divided into three groups based on the analyses. An analysis of molecular variance showed that there was no significant genetic variation between pomegranate genotypes in different locations. Only 2% of the overall genetic variation was due to pomegranates being grown in different locations (locality group differences), while 98% of variation was due to within group differences" [16].

RAPD was used along with the fatty acyl composition of triglycerides in Turkish pomegranate leaves in another analysis [17]. "Seventy-six RAPD decamer primers were used to divide different varieties into three major clusters, while fatty acyl profiles varied between different cultivars. One of the ('Kirli Hanim') had no linoleoyl (called linoleic acid in the paper), while other cultivars did, but in different concentrations. The predominant fatty acyls and their relative concentrations were linoleneate (65 - 71%), palmitoyl (9.4 - 21%), linoeoyl (0.00 – 11.4%), oleoyl (2.4 – 6.2%), stearoyl (0.61 – 4.2%) and myristoyl (0.28 – 9.4%)" [17].

Amplified fragment length polymorphism (AFLP) can also be used to characterize the genetics of pomegranates [18]. "ALFPs are differences (polymorphisms) in the lengths of fragments produced by the hydrolysis of DNA in reactions catalyzed by the restriction endonucleases *Eco*R1 and *Mse*I. Four AFLP primer combinations were used to make 297 DNA fragments, 213 of which were polymorphic. Clustering analysis divided the 19 different pomegranate genotypes into two groups. However, the characteristics of the fruits were quite different between genotypes in the same group. So, genetic analysis is preferred for categorizing the genotypes of pomegranates" [18].

Single nucleotide polymorphisms (SNPs) obtained from transcriptomes revealed genetic diversity in germplasms from pomegranates obtained around the world [19]. "The transcriptome was the collection of all the mRNAs that was produced by leaves, roots, flower parts (petals and reproductive organs) and fruits. The pooled mRNA from these tissues was compared to the mRNA of the peels. The peels did not have the transcripts needed by other tissues for flower and embryo development, pollination, photosynthesis, death, response to extracellular stimulus, and epigenetics. The transcriptome was used to begin a functional annotation of 45817 fully contiguous DNA base sequences (contigs) that were identified with a DNA-protein similarity search using the blastx program. This relied on the similarities of mRNA products to homologous proteins with known functions. Of these, 5943 proteins were seen in the peels as well as other pomegranate tissues, but peels had many more genes from plastids (subcellular organelles in plants)" [19].

That is, in the shotgun approach to DNA sequencing, chromosomal DNA is hydrolyzed into a collection of DNA fragments (300 – 1000 bases). Some of these have overlapping segments that become a consensus or recognized region of DNA. This is a contig.

The proteome analysis also included looking for previously identified proteins from other plants to annotate the pomegranate proteome [19]. "The previously published set of pomegranate contigs was studied with the MIcroSAtellite (MISA) and SciRoKo software searching tools. MISA and SciRoKo found 10330 and 12309 single-sequence repeats (SSRs), respectively. Of these, 7155 were seen using both methods. Most of them were di- and trinucleotide motifs, with the $(AG/TC)_n$ being the most abundant. Next, the DNA of two different pomegranate accessions was analyzed by high-throughput parallel sequencing to look for single nucleotide polymorphisms (SNPs). They found 2336 and 2436 SNPs in the 'Nana' and 'Black' cultivars (accessions), respectively. They had 1728 in common. Also, the genotypes of pomegranates from the germplasm collection at the Newe Ya'ar Research Center were surveyed using 480 of these SNPs. Half of them were highly diverse, with a polymorphism information content (PIC) $\geq$ 0.43. Only 10.7% of the 480 SNPs had minor allele frequencies, so the pomegranate genome is probably quite diverse. Finally, the SNPs were used to divide pomegranates into two large groups. One was from India, China and Iran, but was composed of pomegranates of unknown origin. It was more of an admixture than the other major group, made up of accessions that mainly came from the Mediterranean basin, Central Asia and California" [19].

A much more detailed proteome was recently reported for pomegranate arils [20]. LC with high resolution mass spectrometry (LC-HRMS) was used to analyze an aqueous, pH-buffered extract of powdered arils [20]. Such extracts contain a large dynamic range of concentrations of different proteins [21]. "That is, it can be difficult to see the proteins that are present at the lowest concentrations when the detector is overwhelmed by the predominant proteins. Alternatively, a diverse mixture of peptide ligands can be attached to ProteoMiner beads. It can be used to concentrate the least abundant proteins. There are not enough ligands to bind all the most abundant proteins, so only all small percentage of each of them is bound, while all the low-abundance proteins bind to their specific ligands. Excess high-abundance proteins are washed off, leaving a mixture in which the dynamic range of different protein concentrations is much smaller" [21]. So, the proteome analysis was not quantitative, but it was more nearly complete [20]. It found 1488 proteins, but only six of which that could be identified [20]. Still, this is an excellent beginning and will help guide future studies.

## REFERENCES


[1] Da Silva, J. A. T. et al. *Sci. Hort.* 2013, 160, 85-107.

[2] Holland, D. Hatib, K., Bar-Ya'akov. I. *Hort. Rev.* 2009, 35, 127-191.

[3] Al-Said, F. A. et al. *J. Food Eng.* 2009, 90, 129–134.

[4] Manera, F. J. et al. *Sci. Hort.* 2012, 134, 245-247.

[5] Datacolor. Colorimetric fundamentals, 2009.

[6] Fawole, O. A., Opara, U. L. *Sci. Hort.* 2013, 159, 152-161.

[7] Mayuoni-Kirschenbaum, L., Sadowsky, A., Porat, R. *J. Hort. Sci. Biotechnol.* 2014, 89, 17-22.

[8] Jalikop, S. H. *Fruit Veg. Cereal Sci. Biotechnol.* 2010, 4, 26-34.

[9] Hassan N. A., Gawad, M. H. A. *Amer.-Euras. J. Agric. Environ. Sci.* 2013, 13 (11), 1562-1567.

[10] Donis-Keller, D. et al. *Cell*, 51, 319-337.

[11] Sheidai, M. et al. *Acta Botan. Bras.* 2012, 26(4), 953-965.

[12] Parkinson, J., Blaxter, M. *Meth. Mol. Biol.* 2009, 553, 1-12.

[13] Jian, Z.-H. et al. *J. Genetic.* 2012, 91, 353-358.


[13] Jian, Z.-H. et al. *J. Genetic.* 2012, 91, 353-358.


[14] Alamuti, M. K. et al. *Sci. Hort.* 2012, 146, 104-114.

[15] Orhan, E., et al. *Genetics and Molecular Research: GMR* 13.AOP, 2014.

[16] Noormohammadi, Z. et al. *Austr. J. Crop Sci.* 2012, 6, 268-275.

[17] Ercisli, S. et al. *Biochem. Systematics Ecol.* 2007, 35, 764-769.

[18] Ercisli, S. et al. *Biol. Res.* 2011, 44: 345-350.

[19] Ophir, R. et al. *PLOS One* 2014, 9, e88998.

[20] Capriotti, A. L. et al. *Anal. Bioanal. Chem.* 2013, 405, 9301-9309.

[21] BioRad, ProteoMiner™ Protein Enrichment Kits, Instruction Manual.

*Chapter 3*

# PRODUCTION

## GROWTH AND CULTIVATION

The pomegranate is a shrub that naturally tends to develop multiple trunks, giving it a bushy appearance [1]. "When domesticated, it can grow up to 5 m in height, but wild trees can up to more than 7 m, or even be creeping bushes. Most varieties of pomegranates are deciduous. However, several evergreens can be found in India. Also, various varieties of pomegranates shed their leaves to different extents. Some evergreen cultivars shed their leaves at higher elevations and in colder climates. They are conditionally deciduous" [1].

The young branches produced by recent vegetative growth are usually numerous and thin [1]. "Different varieties produce different colored bark on young branches. The bark can vary from pink to purple, while some are light green with pink-purple spots or stripes. When pomegranates mature, the pink color of the branch starts to disappear. In the second year, the bark will become light gray and continue to darken as the tree matures further. The bark of older trees tends to split, and even becomes detached from the trunk. The wood is light yellow" [1].

Still, young branches sometimes have thorns at their tips that can be seen in the axils in the young blooms [1]. "The young branches resemble quadrangles, but they become round when the branches mature. Young leaves tend to have a reddish color that turns green when mature. In varieties with young pink-purple bark, this color also appears on the sheath and the petiole, on the lower part of the central vein, and in the leaf margins. Leaves resemble oblongs with an obtuse apex and an acuminate base. Mature leaves are green, smooth, and hairless with short petioles. They are usually glossy appearance and contain idioblasts with secretory substances. The leaves are exstipulate, opposed and in pairs with alternate crossing at right angles. Some varieties have three leaves per node arranged at $120^{\circ}$ and may even produce four leaves per node on the same tree" [1].

Pomegranate trees start producing flowers about one month after buds break on newly developed branches, mostly on spurs or short branches [1]. "Solitary flowers can emerge, as can pairs or clusters of them. In most cases, the solitary flowers will appear on spurs along the branches, while the clusters are on the end. In the northern hemisphere, flowering occurs in April and May, but may continue until end of summer, especially in young trees. Such flowers are fertile, but the fruit will not properly mature as trees enter the cooler dormant season in Mediterranean climates. Flowering and the subsequent emergence (setting) of fruits last about one month. During this period, there are three waves of flowering. However, The flowering season of many evergreen cultivars in tropical southern India occurs in three months: June, October, and March, while others produce flowers all year" [1].

In the early stage, the flower can resemble a small pear with a green basal part and red apex or be entirely dark red [1]. "As the flower matures, the sepal becomes orange-red to deep red, depending on the variety. The petals are orange-red or pink and rarely white. As mentioned previously, several pomegranate cultivars from India, Russia, China are ornamental ''double flowered''. These cultivars produce unusually colored petals that are relatively numerous. Some of them are fertile and produce edible fruit, while others are infertile" [1].

There are ten stages of flower development [1]. "In Indian cultivars, flower buds develop between 20 and 27 days. There is a good correlation between the color of the sepals and the final color of the fruit skin. That is, cultivars with deep-red fruit skin will usually have darker red flowers.

Some pomegranate flowers can develop into hermaphrodites that are shaped like a vase or male flowers that are bell shaped. Both types have several hundred stamens. The bell-shaped flower has a poorly developed pistil or none at all, along with atrophied ovaries containing a few infertile ovules. Therefore, the bell shape flower is referred as a male flower and will drop without producing fruit. The vase-shaped flower is fertile with a normal ovary that can develop fruit. The stigma of the hermaphrodite is high in the anthers or above them. This position enables self-pollination as well as pollination by insects. The capacity for producing fruits is determined by the number of vase-shaped flowers. Therefore, cultivars with a higher ratio of vase-shaped to bell-shaped will have a higher yield. The percentage of vase-shaped flowers in the Israeli cultivars ranges from 43% to 66%, while Indian cultivars were 53% to 80%. There is also a third intermediate flower with a short style and a developed ovary that is sometimes fertile" [1].

There are five to eight sepals that are fused in their base [1]. They have a "red fleshy vase shape" [1]. "They do not fall off when the fruit appears, so the fruit develops a prominent calyx. The flower also has five to eight petals, which alternate with the sepals" [1]. They are "separated and have a pink-orange to orange-red color depending on the variety" [1]. "The petals are obovate, very delicate, and slightly wrinkled" [1].

There are also several long stamens in the calyx walls [1]. Often, there are more than 300 stamens per flower. They have orange-red filaments and "yellow bilocular anthers that remain attached to the prominent calyx" [1]. There are also nectaries (gland-like organs that secrete nectar) that are located between the stamens and the ovary base [1, 2]. There are a variable number of carpels, but are usually "eight superimposed in two whorls" [1]. "They form a syncarpic ovary and are arranged in two layers" [1]. The flowers need three to five hours to complete anthesis [1, 3]. The stigma is receptive one day before anthesis and remains that way up to the second day after anthesis [1, 3]. The pomegranate can self-pollinate and also be cross-pollinated by insects, mainly bees [1]. The flowers of Indian, Turkmen, Israeli, and Tunisian pomegranate cultivars can self-pollinate and produce normal fruit [1, 4-6].

"The fruit develops from the ovary and is a fleshy berry" [1]. It is almost round and is "crowned by the prominent calyx" [1]. The apex of the crown can vary from being almost closed to opened wide, depending on the variety and on the stage of ripening. "The fruit is connected to the tree with a short stalk" [1]. The color of the sepals' skin changes from orange-red to green as the fruit develops. "In later stages of fruit maturation, the color will change again until it reaches its final characteristic color as the fruit ripens" [1]. "The external color ranges from yellow, green, or pink overlain with pink to deep red or indigo to fully red, pink or deep purple cover, depending on the variety and stage of ripening [1]. However, there are some unusual cultivars, such as the black pomegranate. The thickness of the skin (leathery exocarp) varies between pomegranate cultivars [1]. "The multi-ovule chambers (locules) are separated by membranous walls (septum) and fleshy mesocarps" [1]. "Usually the lower part of the fruit contains 2 to 3 chambers while its upper part has 6 to 9 chambers" [1]. They are "filled with many seeds (arils)" [1]. "The arils contain a juicy edible layer that develops entirely from outer epidermal cells of the seed, which elongate to a very large extent in a radial direction" [1]. "The color of the edible juicy layer can vary from white to deep red, depending on the variety" [1]. Sometimes, there are several seeds with different colors within an individual pomegranate. The sizes of the arils and hardness of the seeds can vary in different varieties. There are some "seedless varieties" that actually contain soft seeds [1].

"There is no correlation between the outer skin color of the rind and the color of the arils" [1]. The color of the peel (or skin) "does not indicate the extent of ripening degree of the fruit or its readiness for consumption because it can attain its final color long before the arils are fully ripened" [1]. The fruit ripens five to eight months after fruit first appears, depending on the variety [1].

"The pomegranate has a relatively short juvenile period compared to other fruit trees" [1]. When grown from seeds, a small number of seedlings will develop flowers in their first year of growth [1, 7]. They bear fruits in their second year or third year [1]. The color of the fruits will be similar to those of mature pomegranates. "There is a physiological difference between young plants established from seeds (juvenile) and young plants established from cuttings of mature plants" [1]. The time required for seedlings to flower in perennial pomegranates is not always the same as the time needed for young plants that are established from cuttings from mature plants [1].

## IRRIGATION

However, for all of this to happen, pomegranate trees should be grown properly. Even though they do well in hot, dry climates, they need to be irrigated during the dry season to produce maximum yield and optimal fruit quality [1]. "In Israel, irrigation usually starts in late April and lasts throughout the summer, producing yields of 25 to 45 t/ha" [1]. "Drip irrigation is used most commonly in these orchards, although some growers prefer sprinklers (which cause difficulties in weed control)" [1]. Most of the large commercial orchards in Israel, India, and the United States use drip irrigation. "In India and Iran, drip irrigation saved up to 66% of water compared to surface irrigation" [1] as described by Behnia [8] and Chopade and coworkers [9]. "The total amount of water for pomegranate irrigation in Israel for the entire season is 5,000 to 6,000$m^3$/ha, depending on the type of soil and the weather conditions" [1]. "Daily irrigation is practiced during the irrigation season" [1]. "Computerized irrigation yields better results and allows for better control of water quantities and time intervals between successive water applications" [1]. Because of global climate change "and the increasing water shortage experienced in many arid and semiarid regions that are the most suitable regions for pomegranate growth, water availability and irrigation are of major considerations in pomegranate culture" [1].

One group characterized water properties of the 'Mollar de Eche' cultivar in Spain [10]. The water potential of leaves covered by bags, abbreviated by $\Psi_{stem}$, gas exchange, net photosynthetic rate (A) and stromal conductance ($g_s$) were measured under different deficit irrigation regimes. They found that pomegranate trees "have an effective control of the plant water status, by reducing transpiration due to stomatal closure" [10]. They also found that A and $g_s$ are better indicators of water stress than $\Psi_{stem}$. Moreover, "potential pomegranate water needs can be high, particularly during summer" [10]. Finally, the ratio of irrigation plus effective rainfall to reference evapotranspiration can be used to predict the water needs of the trees [10].

In a subsequent study of this Spanish cultivar, $\Psi_{stem}$, midday leaf conductance ($g_l$) and maximum daily trunk shrinkage were evaluated [11]. "Maximum daily trunk shrinkage was found to be the best indicator of how to schedule irrigation in adult trees. In conditions where water was not limited, maximum daily trunk shrinkage could be predicted by the average daily air temperature. So, automated measurements may be able to predict suitable irrigation schedules" [11].

The effect of sustained and regular deficit irrigation of this cultivar was also studied in three seasons [12]. They compared "control irrigated at 100% of crop evapotranspiration (ETc), a sustained deficit irrigation (SDI) where trees were irrigated at 50% of the ETc during the entire season, and three regulated deficit irrigation (RDI) treatments" [12]. "Also, severe water restrictions (25% ETc) were applied during flowering and fruit setting, fruit growth and the final phase of fruit growth and ripening. After 8 to 19 weeks of storage at 5 °C plus 7 days of storage at 20 °C, soluble solids content (SSC), anthocyanins and fruit color were improved by deficit irrigation. So, deficit irrigation can be then used to control the timing of fruit ripening as well as improving the nutritional composition and postharvest stability of fruits" [12].

## Postharvest Treatment

The postharvest biology and technology of pomegranates were reviewed recently [1, 13]. The highlights are summarized here. As mentioned previously, pomegranates are non-climacteric fruits. After the fruits are picked, they continue to respire at a low rate that can remain low when refrigerated [1]. "The respiration rate can be made to increase for a short time by exposing the fruits to ethylene. This does not last long and does not affect the fruit color, TSS, pH or titratable acidity. On the other hand, when the fruits on the trees mature, their titratable acidities decrease, while TSS increases. After being harvested, though, they can lose weight. The peels can be damaged and begin to rot. Also, the fungicides fenhexamid and fluidioxonil can reduce the natural occurrence of the gray mold caused by *Botrytis cinerea*. These fungicides may be more effective when combined with either a wax or a controlled atmosphere of 5 kPascals (kPa) of $O_2$ plus 15 kPa of $CO_2$. In addition, chilling injury can be prevented by pretreating pomegranates with hot (45 °C) water. Heat treatment can also help maintain their nutrient composition and functional properties [1].

It is possible to store pomegranate fruits at 0 – 4.5 °C for several months after being harvested [1]. "Peel damage (husk scald) can be affected by the time of harvest, the temperature, and the oxygen ($O_2$) level. Delaying the harvest can reduce the incidence of husk scalding, as can lower temperatures and less $O_2$. The optimum $O_2$ concentration for minimizing husk scald and preventing the formation of anaerobic volatiles is 5%. However, a combination of 5% $O_2$ and 15% $CO_2$ at 7 °C and 90% to 95% relative humidity may be optimal. Also, weight loss can be minimized while maintaining a good appearance if the fruits are stored in polyethylene bags. A modified atmosphere using special bags with small pores enabled an atmosphere of 5% $CO_2$ and 12 – 14% $O_2$ to develop within the bags that surround the fruits. This reduced weight loss from 7% to 3.5%, reduces scald from 38% to between 2% and 11%, and reduces crown decay after 16 weeks at 6 °C. This procedure can be used to store pomegranates for up to four to five months and keep their commercial quality. Still, antifungal treatment was also recommended before storing freshly harvested fruits and there may be differences with other cultivars that the 'Wonderful' that was tested" [1].

In addition, taste is important when evaluating the quality of pomegranates [14]. "Taste or flavor is a combination of the basic taste, aroma and mouthfeel sensations. The sensation of taste is made up of sweet, sour, bitter, salty and umami attributes. A highly trained descriptive sensory panel evaluated the sensory characteristics of 33 different industrial pomegranate products, including fresh and pasteurized juices, as well as concentrated juice" [14, 15]. "They developed a lexicon of more than 30 sensory attributes for them. Another study found that the overall flavor of pomegranate arils was a due to a combination of various taste (sweet, sour, bitter), aroma (red wine and pomegranate fruity notes) and mouthfeel (astringency, juiciness and seed hardness) sensations. Juice acidity levels, rather than sugar levels, predominantly affect the flavor sensation and preferences of pomegranate arils [14, 15].

Pomegranate juice can also be processed and concentrated by either microwave or conventional heating methods [16]. Microwaves worked better than conventional rotary evaporation in preserving the antioxidant activity and anthocyanin content of the juice [16].

During the commercial processing of whole fruits in the USA, ellagitannins are enriched to a concentration of at least 2 g/L of juice [17]. The residual material can be pressed, treated with a proprietary enzyme solution, concentrated with a membrane system and filtered to produce an extract, POMxl (or POMXL), which is a dietary supplement [18] However, unlike many other extracts described in this book, no organic solvent is used to prepare POMxl. "It was found to be composed of 52% sugars, 2% organic acids, 2% ash and 130 mg/mL (13%) total phenolics, expressed as gallic acid equivalents (GAE). The phenolics were mostly hydrolyzable tannins (ellagitannins), ellagic acid, anthocyanins and their glycosides. POMxl was also used to prepare a pomegranate polyphenol extract (POMxp), which was 93% phenolics, with 3.2% sugars, 1.9% organic acids, 2.5% ash, 5.0% protein and 1.2% moisture. They also described extracts obtained from arils (POMa), flowers (POMf), and seed oil (POMo). The POMa was 79% carbohydrates, 7% protein, 2.5% ash, 1.4% total phenolics, 0.2% fat and 10% moisture. The POMf was 30.2% dietary fiber, 7.7% sugars, 2.0% sugar alcohols, 22% other carbohydrates, 5.6% ash, 9.4% protein, 0.9% fat and 5.7% moisture. The POMo had no phenolics since it was prepared by extraction with hexane, followed by refining, bleaching and deodorizing" [18]. The authors said that it contained fatty acids [18], but a hexane extract of seed oil would actually contain triglycerides, with fatty acyls attached to a glyceride backbone. Still, the triglycerides are good sources of dietary fatty acids once they are digested. Also, the primary fatty acid produced by hydrolyzing the triglycerides was punicic acid (64.3%) [17, 18]. A dose of two POMx capsules weighs 2.1 g and contains 1.8 g of carbohydrates, 5.86 mg of fat, 72.5 mg of protein, 113 mg of ash, 116 mg of moisture 372 mg ellagic acid and 1505 mg of total phenolics [19].

So, there are a variety of products that can be made from pomegranates.

## REFERENCES


[1] Holland, D. Hatib, K., Bar-Ya'akov. I. *Hort. Rev*. 2009, 35, 127-191.

[2] Fahan, A. The leaf. pp. 171–212, The flower, pp. 321–394, The seed, pp. 419-430, in Plant anatomy. Hakkibutz Hameuhad Publi., Jerusalem, 1976.

[3] Josan, J. S., Jawanda, J. S., Uppal, D. K. *Punjab Hort. J*. 1979, 19, 66-70.

[4] Nalawadi, U. G. et al. Studies on the floral biology of pomegranate (*Punica granatum* L.) Mysore *J. Agr. Sci*. 1973, 7, 213–225.

[5] Mars, M., Marrakchi, M. *Genet. Resour. Crop Evol*. 1999, 46, 461–476.

[6] Karale, A. R., Sanghavi, K. U., Patil, A. V. *Res. Bul. Marathwada Agr. Univ*. 1979, 3, 57–59.

[7] Terakami, S. et al. *Plant Cell Rep*. 2007, 26, 1243–1251.

[8] Behnia, A. Comparison of different irrigation methods for pomegranate orchards in Iran. Irrigation under conditions of water scarcity. 17th Int. Congr. Irrigation and Drainage, 13–17 Sept. Granada, Spain, 1C:207–217.

[9] Chopade, S. Q. et al. *Adv.Hort. Forest*. 2001, 8, 53–59.

[10] Intrigliolo, D. S. et al. *Agr. Water Manage*. 2011, 98, 691-696.

[11] Galindo, A. et al. *Agr. Forest Meteorol*. 2013, 180, 58-65.

[12] Laribi, A. I. et al. Effect of sustained and regulated deficit irrigation on fruit quality of pomegranate cv. 'Mollar de Elche' at harvest and during cold storage. *Agric. Water Management* 2013, 125, 61-70.

[13] Kader, A. A. Postharvest biology and technology of pomegranates. pp. 211–220, in *Pomegranates: Ancient Roots to Modern Medicine*. CRC Press Taylor & Francis Group, Boca Raton, FL, 2006.

[14] Mayuoni-Kirshinbaum, L., Porat, R. *J. Sci. Food Agric*. 2014, 94, 21-27.

[15] Koppel, K., Chambers IV, E. *J. Sens. Stud*. 2010, 25, 819-837.

[16] Fazaeli, M., Yousefi, S., Emam-Djomeh, Z. *Food Res. Intl*. 2013, 50, 568-573.

[17] Bialonska, D. et al. *J. Agric. Food Chem*. 2009, 57, 8344–9.

[18] Aviram, M. et al. *J. Agric. Food Chem*. 2008, 56, 1148-1157.

[19] Basu, A. et al. *J. Nutr. Metab*. 2013, Article ID 708381, 7 pages.

*Chapter 4*

# CHEMICAL COMPOSITION

There are many bioactive phytochemicals in pomegranates. This includes sterols and terpenoids in the seeds, bark, and leaves; alkaloids in the bark and leaves; triglycerides in seed oil; derivatives of gallic acid in the leaves; organic acids in the juice; flavonols in the rind, fruit, bark, and leaves; anthocyanins and anthocyanidins in the juice and rind; and catechin and procyanidins in rind and juice [1]. According to the US Department of Agriculture (USDA) national nutritional database, raw pomegranates contain 77.93% water along with 1.67%, 1.17%, 18.7%, 4.0% and 13.67% protein, total lipids, carbohydrates, dietary fiber, and total sugars, respectively [2]. They also contain 10, 0.30, 12, 36, 236, 3 and 0.35 mg/100 g of calcium, iron, magnesium, phosphorus, potassium, sodium and zinc, respectively [2]. They also contain 10.2, 0.067, 0.053, 0.293, 0.075 and 0.60 mg/100 g of vitamin C, riboflavin, niacin, vitamin B-6 and vitamin E (α-tocopherol), respectively [2]. They also contain 38, and 16.4 μg/100 g of oleoyl (oleic acid as part of triglycerides) and vitamin K, respectively [2]. The USDA database also claims that pomegranates contain 0.12, 0.093 and 0.079% saturated, monounsaturated and polyunsaturated fatty acids [2].

Actually, they do not contain any fatty acids, unless the oil in the seeds has turned rancid. They contain triglycerides that have fatty acyls and not fatty acids covalently attached to glycerol. Fortunately, few physicians or consumers read articles published in journals that focus on food chemistry. Instead, physicians read medical journals and consumers read labels on the packages containing foods, juices and dietary supplements. They almost always use the proper terms, saturated fats, monounsaturated fats, polyunsaturated fats, *trans* fats and omega-3 fats (or just simply omega-3s).

Regardless of the nomenclature used, the concentrations and amounts of these compounds in the pomegranate change during its development and fruit maturation. Some of them can be measured accurately using standard methods. For example, there are standard indices for measuring color. Water or moisture content can be determined by loss of weight upon drying. Minerals are solubilized by a thorough acid digestion and sugars like fructose and glucose dissolve readily in water. For lipophilic (hydrophobic) triglycerides, the standard method is to extract liquid samples that contain water (like juices) with twice their volume of 2:1 chloroform:methanol (v/v). The methanol ($CH_3OH$) helps break apart tight non-covalent bonds between proteins and triglycerides, lipids and other lipophilic compounds, which enter the chloroform ($CHCl_3$) phase [3]. Still, for seed oils, hexane is often preferred for solubilizing fats, but $CHCl_3$ is more convenient for NMR (nuclear magnetic resonance) analysis.

However, the actual concentrations of organic acids, flavonols, anthocyanins, anthocyanidins, procyanidins and other hydrophilic organic nutrients may be underestimated because of inadequate extraction. That is, the first step in analyzing samples is to extract the analytes. However, there is no one standard method for extracting hydrophilic compounds from solid samples. This is not a concern for juices, which simply can be diluted into water [4] or $CH_3OH$ [5], mixed with acetone [6] or just centrifuged to remove insoluble solids [7, 8]. After dilution, total phenolics are measured by a standard assay using a Folin-Ciocalteu colorimetric reagent [4-7, 9, 10], while total anthocyanins are measured by the intensity of their color, or visible absorption spectrum [4-7, 10]. Total flavonols or flavonoids can be quantified by forming a colored complex with aluminum ($Al^{3+}$) that absorbs at 430 nm [10]. Individual phenolics and anthocyanins can be separated by liquid chromatography and detected by UV-visible absorption [7] and/or mass spectrometry [11].

On the other hand, solid samples, such as whole fruits, arils, peels, seeds and dietary supplements need to be extracted with water and/or organic solvents [12-16]. Most studies shake or sonicate samples with methanol and/or water at room temperature and pressure, but it is unlikely that all the phenolic compounds can be extracted this way. Many of them are packed tightly inside of plant cells that are surrounded by rigid cell walls. They can be tightly bound to other compounds in the cells. Few analysts do a detailed study to find the best extraction method and few journals will publish such studies. Still, one clever study dissolved 100 mg of dietary supplement samples in 100 mL dimethylsulfoxide (DMSO) [17]. This solvent will rapidly penetrate cell walls in plants and even skin in humans. It will take all the solutes that are dissolved in it directly into the blood stream. So, even though DMSO itself has very low toxicity, it can be very dangerous if it contains dissolved toxins.

Another option is to use dry methanol at elevated temperature (100 $^{o}C$) and pressure (10 MPascal or 100 atmospheres) in a pressurized container, as part of an accelerated solvent extractor, or ASE [18]. At high temperature and pressure, $CH_3OH$ can penetrate cell walls, explode the cells, break non-covalent bonds between analytes and solubilize them. The ASE has been shown to be able to extract much more material than other methods from freeze-dried fruits, such as noni (*Morinda citrifolia*) [19]. Moreover, dry methanol has no water in it, so easily hydrolyzable anthocyanins can be preserved. So, the actual concentrations and contents of soluble solids, phenolic compounds, flavonols and anthocyanins may be underestimated if they are not all extracted completely from solid samples. Still, if the same extraction method is used in the analysis all samples, the relative concentrations of analytes may be accurate. So, it can be possible to observe changes in analyte concentrations that occur as pomegranate fruits mature.

One study found that the concentrations of total phenolics, antioxidant activity and hydrolyzable tannins were reduced in the peels during maturation, while the anthocyanin level increased [4]. The results show that the sugar content in the aril juice increased while the levels of acidity and of citric acid decreased. However, even though the antioxidant and total phenolic contents significantly decreased in 'Rosh-Hapered', it didn't happen in the 'Wonderful' variety. On the other hand, the anthocyanin concentrations increased in 'Wonderful' but did not change in 'Rosh-Hapered' [4].

In another study, three Spanish cultivars, 'Mollar de Eche', 'Borde de Abatera' and 'Piñon Tierno de Ojos' were analyzed at different stages of maturity [20]. The main quality parameters were total soluble solids, titratable acidity, pH and maturity index. The concentrations of organic acids, sugars, proline (an amino acid), total phenolics, antioxidant activity and color were determined. The position of the fruit on the tree only affected its color – not the other parameters. It should be noted that the standard method for determining total soluble solids uses refractometry, which actually measures the refractive index due to the presence of sugars, such as sucrose and fructose. The unit of measurement is $^{o}$Brix. It is assumed that the $^{o}$Brix is equal to the percent sugar in the sample, if it is measured at or about 20 $^{o}$C. This may be valid for fruits that have relatively high sugar contents, but there are esters of sugars and fatty acyls that can be present at appreciable levels in some fruits. They require high temperature and pressure extraction with methanol and an ASE to be completely solubilized and identified by NMR [19].

Still, refractometry is useful for estimating the total concentration of water-soluble sugars. It was used to find that the total concentration of soluble sugars was 14.8 to 16.5 $^{o}$Brix for the three cultivars when mature [20]. The total concentrations of sugars were 108 to 133 g/L, with glucose and fructose being the most abundant when analyzed by a different method. Sugar concentrations decreased slightly as the fruits matured. Total acidity was measured by titration. It ranged from 2.29 to 21.35 g/L when expressed as citric acid equivalents. The individual acids were separated and quantified by high performance liquid chromatography (HPLC). Citric, malic and quinic acids were the predominant organic acids. The concentration of proline ranged from 78 to 89 mg/L in mature fruits. It increased as the fruits matured. Total phenolics ranged from 2674 to 3876 in ripe fruits and decreased as the fruits matured [20].

Chinese pomegranate peels at different stages of maturity were analyzed for anthocyanins, which were solubilized by sonication in methanol containing 0.1% HCl [21]. Even though this may not have solubilized all the anthocyanins, the results did indicate that "the anthocyanins varied with different cultivars and with different stages of maturity. Moreover, the major anthocyanins were cyanidin-3-*O*-glucoside (Cy3G), delphinidin-3-*O*-glucoside (Dp3G), delphinidin-3,5-*O*-glucoside (Dp3,5dG), cyanidin-3,5-*O*-glucoside (Cy3,5dG), pelagronidin-3-*O*-glucoside (Pg3G) and pelagronidin-3,5-*O*-glucoside (Pg3,5dG). Concentrations of Cy3G ranged from 1.56 to 53.5 mg per 100 g of fresh fruit" [21].

Five different stages of maturity were defined for pomegranate fruits from the 'Ruby' cultivar [22]. "This was based on fruit mass, size, juiciness, color, total soluble solids, pH, titratable acidity, individual organic acids, sugars and phenolic compounds. Principal component statistical analysis showed that the colors of the arils and peels and the ratio of sugars to organic acids were the most important factors for measuring the maturity and ripeness of the fruits. Ripe fruits had an average mass of 321.5 g, a diameter of 8.44 cm, length of 7.48 cm, a pH of 3.28, titratable acidity of 0.31%, 188 mg tartaric acid per 100 mL, soluble solids of 15.2 $^{o}$Brix, 613 and 675 mg of glucose and fructose, respectively, per 100 mL, and a juiciness of 49.3 mL per 100 g of arils" [22].

The physicochemical properties of the 'Taifi' variety (or cultivar) from Saudi Arabia were also determined as fruits matured [23]. "The fresh, ripe fruits had an average mass of 191 g, of which 69 g were from the skin (peel). The average length was 6.55 cm and diameter was 3.67 cm. The edible portion of the fruit was 57.5% of the total fruit weight. It was 63.6% juice and 36.2 % seeds. Fresh juice contained 84.57% water, 14.1% sugar, 1.05% protein and 0.33% ash. Total protein, ascorbic acid, fat and phenolic compounds in the seeds were 4.06, 0.23, 0.15 and 2.92%, respectively. The pH of the fruit increased as the fruits matured. The pH of the juice was 3.48 when the fruits were fully ripe. Ripe fruits were significantly less acidic than green, unripe and half-ripe fruits. Ripe fruit juice had 7.51 g of glucose and 6.56 fructose per 100 g of fresh, mature fruit. Polyphenol concentrations were lower in full-ripe fruits than unripe. The concentrations of K, Na, Mg and Ca were 243, 95.7, 11.9 and 59.3 mg per 100 g of fresh fruit" [23].

The changes in organic acids, sugars, ascorbic acid and total anthocyanins during the maturation of the South African 'Bhagwa'cultivar were determined [24]. "The sugars, ascorbic acid and total anthocyanins increased, while the total phenolics, titratable acidity and organic acids decreased. It was quite interesting that this cultivar contained not only the expected acids (citric, malic and ascorbic), but also tartaric acid (at 250-380 mg/100 mL juice)" [24], since tartaric acid is not usually seen at anything above trace levels in other cultivars [25, 26].

The biochemical composition and antioxidant activity of the arils of the Indian cultivar 'Ganesh' were also found to change as its fruits mature [27]. "The total soluble solids, total sugars and reducing sugars increased duting the first 100 days of fruit development. Ascorbic acid and total phenolics decreased from days 20 to 100. There was also an initial decrease in protein content, followed by an increase from days 20 to 100. Even though the antioxidant activity decreased from days 20 to 60, it increased in the later stages of development, due to an increase in total anthocyanins. Finally, the anthocyanin content and total acidity decreased from 100 days onwards, which caused browning of the arils" [27].

There are also changes in chemical composition and color during the processing of pomegranates [6]. When the juice is clarified and pasteurized, some of the anthocyanins can be destroyed by hydrolysis. There can also be changes in pH, titratable acidity and total soluble solids, as seen in some Turkish cultivars [6]. Before processing, the concentrations of total anthocyanins in the juice from the sacs and juice from the whole fruits were 382 and 291 mg/L, based on HPLC analysis, compared to 364 and 365 mg/L by the visible spectrophotometric method.

That is, HPLC can separate and detect the individual anthocyanins, while spectrophotometry measures the blue color that is due to all of the anthocyanins put together. Neither method is perfect. The HPLC method measures the peak areas produced by each of the six anthocyanins. The same ones that were in Chinese pomegranates [21] (Cy3G, Dp3G, Pg3G, Cy3,5dG, Dp3,3dG and Pg3,5dG) were found in the Turkish cultivars. However, they were quantified by comparing the peak areas of standards to those seen in the juice samples [28]. There was no Dp3,5G standard, so "it was quantified based on the response due to the Cy3G standard". However, none of the standards used (Cy3G, Cy3,5dG, Dp3G, Pg3G and Pg3,5dG) are available in sufficient quantity to ascertain their purities. Instead, vendors can provide a certificate of analysis that includes an HPLC chromatogram that shows only one peak. However, when anthocyanin standards are purified, water, acetic acid and/or NaCl are used in the purification process. Traces of them may remain after the purification is done. However, none of them are detected by HPLC when absorbance at 520 nm (or any wavelength in the visible region) is used for detection. As a result, very different absorbance spectra and molar absorptivities have been reported for the most readily available standard, Cy3G" [28].

The visible spectrophotometric method for determining total anthocyanin concentration measures the difference in absorbance at pH 1.0 and 4.5. The Turkish workers used a molar extinction coefficient of 26900 $M^{-1}cm^{-1}$ at 512 nm ($\varepsilon_{512}$) to calculate the total anthocyanins

[6]. This assumes that all of the anthocyanins have the same $\varepsilon_{512}$ as Cy3G. This may not be accurate, but if the same method is used to analyze all samples, the results can be compared. So, both methods showed that clarification and pasteurization can decrease the concentrations of both the total anthocyanin and individual anthocyanins. Still, it took HPLC to establish that there was more Cy3,5dG in the Turkish 'Hicaz' variety than other anthocyanins. The Turkish workers also found that cold clarification using just gelatin was sufficient, if done carefully. There was no need to remove pectin [6]. Also, clarification and pasteurization can decrease total anthocyanin content, so both processes should be optimized to limit this loss [6].

In addition to anthocyanins, there are also other phenolic compounds, most notably gallic acid, ellagic acid, punicalagin A and punicalagin B [14]. "The punicalagins are unique to pomegranates. All four of these compounds were quantified in several pomegranate products by means of HPLC with UV-Visible spectrophotometric detection. POM Wonderful pomegranate juice had 69.6, 152, 179 and 113 mg/L of gallic acid, ellagic acid, punicalagin A and punicalagin B, respectively. Other juices contained different concentrations of these compounds. A sample that was called both a peel extract and a marc extract had 32.6, 40.1, 43.2 and 98.1 mg/L of these compounds. The marc extract was obtained from POM Wonderful LLC (Del Rey, CA, USA). The extract was prepared by stirring a mixture of water and marc (50:1) at room temperature for 20-60 mins" [14]. The marc material is a by-product of juice production [15]. "About 3300 tons of it wass obtaineded from 20.5 tons of Californian pomegranates that produced about 332 L per ton of fruit. The marc contained about 78% peels and 22% seeds" [15]. It is quite possible that much more could have been extracted using pressurized liquid extraction with either water or ethanol. So, the actual concentrations of these healthy phenolic compounds could actually be much higher in the marc material.

So, punicalagin A and punicalagin B can be used along with punicalin and ellagic acid to try to verify the authenticity of 27 dietary supplements made from pomegranates [17]. Only five of them had the typical profile of pomegranate tannins when analyzed by HPLC, 17 had ellagic acid with little or no detectable pomegranate tannins, and five others had no detectable tannins or ellagic acid [17]. There are commercial laboratories that analyze pomegranate juice and other dietary supplements from various vendors. For a nominal fee, consumers can access the reports and find out which products have been analyzed and which ones appeared to be fraudulent. The results of analyses of juices can have legal consequences. In June, 2014 the United States Supreme Court ruled that companies can be sued for false advertising when they label a mixture of juices as containing pomegranate juice if it is present at a concentration of only 0.3%.

So, the international community has established some standards for verifying the authenticity of pomegranate juice [26, 27]. First, "it should have the same six anthocyanins described above, along with the punicalagins. The total sugar concentration (often called total soluble solids) should be about 16 $^{o}$Brix. There should be mannitol at >3 g/L. The ratios of glucose to mannitol should be from 4-15 and glucose to fructose should be 0.8-1.0. To adjust the astringency of the juice or peel extract, nonpomegranate sugars are often added" [26]. The presence of sucrose could indicate adulteration with peach juice [27]. So, there should not be any sucrose in pomegranate juice [26, 27]. Also, there should be no maltose or tartaric acid [26]. If the amino acid proline is present at >25 mg/L, or if there is any tartaric acid, grape products have been added. Malic acid at >0.1 g/100 mL indicates adulteration with apple, pear, grape, cherry, plum, or aronia juice [26]. Finally, adulteration of pomegranate juice with grape juice causes significant increases in acetic acid, isoamyl butyrate and especially 1-hexanol and linalool, while adulteration with peach juice increases the concentrations of butyl acetate, isobutyl butyrate, benzyl acetate and especially isoamyl butyrate [27].

There are other important parameters. A sample of the 'Wonderful' variety of juice had a pH of 2.98, a total acidity of 17.5 g citric acid/L, and a sugar content of 17.8 $^{o}$Brix [29]. 'Coupage' juice had a pH of 3.01, a total acidity of 10.0 g citric acid/L, and a sugar content of 17.4 $^{o}$Brix. 'Mollar de Elche' juice had a pH of 3.52, a total acidity of 2.93 g citric acid/L, and a sugar content of 16.9 $^{o}$Brix [30].

The flavor of 'Wonderful' pomegranate arils was evaluated by a panel of trained tasters and analysis of volatiles by gas chromatography coupled to mass spectrometry (GC-MS) [31]. "The flavor was acceptable during the first 12 weeks of storage at 7 $^{o}$C, but decreased after 16-20 weeks. There was a high correlation between overripe and off-flavor odors and the concentrations of ethanol and ethyl acetate, which are products of fermentation. The presence of volatile sesquiterpenes also damaged the flavor" [31].

So, pomegranate juice and wine should have characteristic mixtures of volatile compounds [32]. "The terpene called limonene is the most abundant volatile compound in juice, while ethyl octanote is a marker for wine. There were some differences in volatiles in different varieties of pomegranates. Volatiles also dominate the taste and other sensory profiles, which can be identified by qualified taste testers" [32].

Wines made from the 'Wonderful', 'Coupage' and 'Mollar de Elche' had pH values of 3.12, 3.14, and 3.35, respectively [29]. "Their total acidities were 20.2, 12.3, and 4.6 g citric acid/L; volatile acidities were 0.33, 0.36, and 0.26 g/L for 'Wonderful', 'Coupage' and 'Mollar de Elche', respectively. Finally, alcohol contents were 8.3%, 8.7%, and 9.1% v/v for 'Wonderful', 'Coupage' and 'Mollar de Elche', respectively. Also, the concentration of malic acid decreased during fermentation, while acetic acid emerged and its concentration increased during fermentation. The concentrations of anthocyanins, ellagic acid and total phenolics also decreased, while gallic acid increased. Still, the antioxidant capacities of the wines remained rather high" [29].

The antioxidant capacities, total phenolics and total anthocyanins of pomegranate juices obtained from eight different cultivars from Turkey were measured [33]. "The antioxidant capacities were measured by two different methods. One of them measured the ability to scavenge a free radical (DPPH) in aqueous solution and the other measured the anti-lipid peroxidative activity (ALPA). The 'Izmir 8' variety had the highest antioxidant activities. The ability to scavenge DPPH was 418 mg/100 mL of juice and the ALPA was 93.5%. It also had a total anthocyanin content of 8.2 mg/100 mL (based on Cy3G) and 309 mg/100 mL total phenolics" [33].

The arils, juices and wines are not the only products obtained from pomegranates. The peels, seeds, leaves and flowers are also used. One study extracted 15 g of dried, powdered peels from five different Egyptian cultivars two different ways. They used either 150 mL 80% methanol (and 20% water) or 100% water, presumably after shaking the mixture [34]. "The concentrations of total phenolics that they reported ranged from 179 to 246 mg per g of peels in the extracts prepared using 100% water. Less was extracted using 80% methanol. Total flavonols ranged from 107 to 137 mg/g, with the 80% methanol solubilizing more than 100% water. The proanthocyanidin contents were 85 to 340 μg/g, while total hydrolyzable tannins ranged from 519 to 730 μg/g. Water extracted more hydrolyzable tannins, but 80% methanol extracted more proanthocyanidins" [34]. However, it is unlikely that either solvent extracted all of them. Pressurized liquid extraction will almost certainly extract much more. Dry methanol at 100 °C and 10 MPascal pressure might be the best choice, since anthocyanins and proanthocyanidins are susceptible to hydrolysis.

The phenolic compounds in four Tunisian cultivars were characterized by matrix assisted laser desorption, time of flight mass spectrometry (MALDI-TOF MS) [13]. "About 0.5 g of dried, powdered peels were extracted three times with 10 mL of 80% methanol using a homogenizer. The total phenolics ranged from 100 to 181 mg per g of dry weight (mg/g DW). There were 1.2 to 3.1 mg/g DW of condensed tannins and 422 to 505 mg/g DW of hydrolyzable tannins. Also, several flavonoid oligomers were seen, with some being pentamers and hexamers" [13].

Pomegranate peels obtained from Pakistan were found to have about 4% water, 5% ash, 21% fiber, 9.4% fat, 0.7% total soluble solids, 4.86% total acidity, 31% total sugars, 30 % reducing sugars, 1% non-reducing sugars, 1.4% nitrogen, 8.7% protein and a pH of 3.75 [35]. "It also had 1100, 1000, 60.5, 4.5 and 4.0 μg/g sodium, potassium, iron, manganese and zinc, respectively. The water, fiber, fat, nitrogen, ash and minerals contents were measured accurately, but the others probably were not. The article did not say how the samples were extracted before analyzing them for total soluble solids, total acidity, sugars and protein. Moreover, pH is only applicable to aqueous solutions. So, it can be measured accurately in juices and aqueous extracts, but not in peels that have only 4% water. Instead, it is quite likely that the aqueous extract that was prepared in this study for determining antimicrobial and antifungal activities was the sample whose pH was measured. Also, it is possible that the total soluble solids, total acidity, sugars and protein concentrations were measured on the aqueous extract that was prepared by mixing 50 g of powdered sample with 200 mL distilled water and agitating the mixture for 6 hrs, followed by filtration. It is also quite unclear how the investigators found only 0.7% total soluble solids, but 31% total sugars" [35].

Pomegranate peels from 12 different Tunisia cultivars were analyzed for dietary fiber, water and oil-holding properties, soluble phenols and antioxidant activity [36]. "The total dietary fiber ranged from 33-62%. It was made up of 16-23, 21-42, 14-23 and 17-20 g/100 g cellulose, Klason lignin, uronic acid and total neutral sugars, respectively. Arabinose and xylose were the most abundant neutral sugars. The ratio of insoluble to soluble DF was around 1, reflecting its balanced composition. The water- and oil-holding capacities ranged from 2.31-3.53 and 2.80-4.05 mL/g, respectively. Total concentrations of phenolic compounds ranged from 205 to 276 mg/g, but they were extracted using 80% ethanol at room temperature and pressure" [36], so there may be a much higher concentration of phenolic compounds than this.

The seeds can be used to prepare a nutritious oil. Even though most investigators used gas chromatography to analyze hydrolyzed oil, at least one group used MALDI-TOF MS so they could see the triglycerides [37]. "Still, they did talk about the supposed fatty acid profile. As mentioned previously, the oil obtained from vegetables and some fruits (like olives, açaí and pomegranate seeds) does not contain any fatty acids unless it has turned rancid. They contain mostly triglycerides, with some samples having small amounts of mono- or diglycerides. Still, this article did find that unsaturated fats with 18 carbons and three carbon-carbon double bonds were the predominant fatty acyls in the triglycerides. This fatty acyl is abbreviated 18:3. There were four different isomers, with punicic acid [sic] (18:3: 9-*cis*, 11-*trans*, 13-*cis*) being the most abundant. So, even though the investigators reported that linolenic acid [sic] was the predominant fatty acid [sic], it was actually punicic acid [sic], which they said was part of the linolenic acid fraction. In addition, an unusually high level of phytosterols were found (4089-6205 mg/kg). The major ones were β-sitosterol, campesterol and stigmasterol" [37].

Another group analyzed 25 different varieties of Iranian pomegranate seeds [38]. "They found that the seeds contained 66.3 to 193 g/Kg DW of oil. They reported that the predominant fatty acid [sic] was linolenic acid (actually linoleoyl), followed by linoleic, oleic, stearic and palmitic acids [sic]. That is, the oil contains primarily unsaturated fats, with the exception of one variety called Gorche Shahvar Yazdi" [38]. As mentioned previously, fruit and vegetable oils do not contain fatty acids unless they have turned rancid. Still, a clever sales person from another country or competing company could claim that this paper "proves" that Iranian pomegranate oil is inferior, due to the presence of fatty acids.

Similarly, another group reported finding fatty acids in the seeds of seven clones of lesser known sweet Spanish pomegranates [39]. "The oil content of the seeds ranged from 63 to 122 g/kg DW. They reported that the predominant fatty acid [sic] was linolenic acid, followed by linoleic acid, oleic and palmitoleic acid. The ratio of saturated to unsaturated fats was 0.04 to 0.35" [39]. Again, a clever sales person could say that this paper "proves" that these clones are inferior due to being rancid.

In still another paper, a commercially available pomegranate seed oil contained primarily punicic acid [sic], but not jacaric acid [40]. They did admit that the oil was made up of triglycerides, but said that they contained fatty acids. Fortunately, they did not identify the source or name of the pomegranate seed oil, so competing vendors cannot criticize it for being rancid.

In contrast to the seeds, the leaves don't produce an oil, but they do contain phenolics, flavonoids and alkaloids [41]. "They are used in China to make teas that are widely consumed. The total concentrations of phenolics and flavonoids decreased in the early stages of leaf growth, and then increased gradually until the end of September. At the same time, the concentrations of total alkaloids increased during leaves grew and developed. These results suggest that the optimum picking time for the leaves is before May and bioactive compounds should be extracted from pomegranate leaves after August in China" [41].

Another group looked at leaves from Tunisian pomegranates [42]. "They reported finding 8.8 – 127.3, 1.2 – 76.9, 63.7 – 261, and 0.41 – 3.73 mg/g DW of total phenolics, flavonoids, tannins and anthocyanins, respectively. The different concentrations do not reflect differences in varieties or cultivars, but rather different solvents that were used for extraction. They extracted dried leaves at room temperature and pressure with five different solvents. First, they mixed 10 g of leaves with 100 mL of hexane for 16 hrs with frequent agitation. This was repeated twice. After filtering off the hexane, the leaves that remained were extracted the same way with dichloromethane, ethyl acetate, ethanol and finally methanol. Hexane solubilized the fewest phenolics and no flavonoids, but the most tannins and anthocyanins" [42]. Still, it is quite likely that pressurized solvent extraction with methanol or ethanol at elevated temperature would extract much more of all of these classes of compounds in one easy step.

Flowers from Iranian pomegranates have also been analyzed [43]. "The 'Ghojagn' cultivar had the highest concentration of total phenolics (26 mg/g), but the 'Danesiah' cultivar had the most total flavonoids (23 mg/g). The 'Rabbab' cultivar had the most tannins (2.03%). Finally, the 'Ghojagn' cultivar had the highest antioxidant capacity. For all but the tannin analysis, dried flowers were extracted with 80% methanol at room temperature. Dried flowers were extracted with water when preparing them for tannin analysis" [43]. So, like so many other studies, this one probably seriously underestimated the concentrations.

In another study, flowers from Iranian pomegranates were analyzed for their concentrations of total phenolics, tannins and antioxidant capacity [44]. "The concentration of total phenolics ranged from 15.2 to 25.9 mg GAE/g DW for six different cultivars. The 'Ghojagh' cultivar had the highest. The total tannins were between 1.06 to 2.03%, with the 'Rabbab' cultivar being the highest. The total flavonoids were 11.5 to 23.1 catechin equivalents per gram, with the 'Danesiah' cultivar being the highest" [44]. The ferric reducing antioxidant power (FRAP) ranged from 123 to 452 $Fe^{2+}$ equivalents/g DW, with the 'Ghojah' cultivar being the highest.. The antioxidant capacities measured by the DPPH assay were between 97.6 to 116 mg vitamin E equivalents/g DW, with the 'Malas' cultivar being the highest.

Pomegranate flowers also contain oleanolic and ursolic acid [45]. "One of the groups took the time to optimize the extraction method. They found that ultrasonication of dried flowers for 50 min at 40 °C with 90% ethanol at a ratio of 20:1 (v/w) for solvent to flowers was best. They found 5.31 and 9.11 mg/g DW of oleanolic acid and ursolic acid, respectively" [45]. It should be noted that oleanolic acid is quite different than oleic acid or oleoyl (as part of a triglyceride). Oleanolic acid is a pentacyclic triterpene (Figure 1), while oleic acid is a fatty acid (Figure 2).

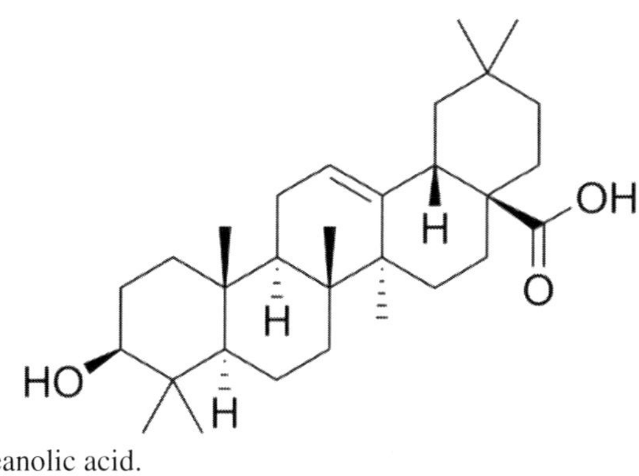


Figure 1. Structure of oleanolic acid.

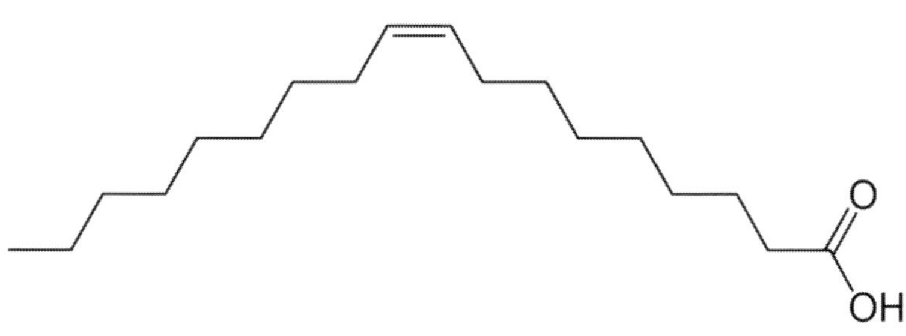


Figure 2. Structure of oleic acid.

Finally, pomegranates have proteins. There are about 120 g of proteins per kg of seeds [46]. In a recent study, 1488 different proteins were found in the arils of the Italian cultivar 'Dente de cavallo' [46]. Only six of them were already in international databases [46].

In conclusion, many studies have been published that identify the metals and organic compounds that are present in pomegranate fruits, juices, and other parts. The metals contents are probably accurate, since the samples are always digested completely with concentrated acids. Also, the fat contents of the seed oils are also accurate since fats are completely extracted by hexane or chloroform. However, investigators often mistakenly report finding fatty acids in the seed oils, even though they actually contain fatty acyls as part of triglycerides. As mentioned in the introduction, this is a common mistake in the literature on food chemistry. Also, the concentrations of hydrophilic compounds, such as phenolic compounds (including anthocyanins, proanthocyanidins, flavonoids and tannins) are probably underestimated due to inadequate extraction at room temperature and pressure. It would be better to use pressurized liquid extraction. Finally, the yields of commercially prepared extracts from pomegranates could be greatly improved, especially if dry ethanol or pressurized water were used.

## References

[1] Holland, D. Hatib, K., Bar-Ya'akov. I. *Hort. Rev.* 2009, 35, 127-191.

[2] US Department of Agriculture, National Nutritional Database for Standard Reference, Basic Report 09286, Pomegranate. http://ndb.nal.usda.gov/ndb/foods/show/2434.

[3] Bligh, E. G., Dyer, W. J. *Can. J. Biochem. Physiol.* 1959, 37, 911-917.

[4] Shwartz, E. et al. Changes in chemical constituents during the maturation and ripening of two commercially important pomegranate accessions. *Food Chem.* 2009, 115, 965-973.

[5] Al-Maiman, S. A., Ahmad, D. Changes in physical and chemical properties during pomegranate (*Punica granatum* L.) fruit maturation. *Food Chem.* 2002, 76, 437-441.

[6] Turfan, O., Turkyilmaz, M., Yemis, O., Ozkan, M. Anthocyanin and colour changes during processing of pomegranate (*Punica granatum* L., cv. Hicaznar) juice from sacs and whole fruit. *Food Chem.* 2011, 129, 1644-1651.

[7] Vegara, S. et al. *Food Chem.* 2014, 147, 203-208.

[8] Gómez-Caravaca, A. M. et al. Determination of the major phenolic compounds in pomegranate juices by HPLC−DAD−ESI-MS. *J. Agr. Food Chem.* 2013, 61, 5328-5337.

[9] Singleton, V. L. Meth. Enz. 1999, 299, 152-178.

[10] Hmid, I. et al. Comparative study of phenolic compounds and their antioxidant attributes of eighteen pomegranate (*Punica granatum* L.) cultivars grown in Morocco. *Arabian J. Chem.* 2013.

[11] Gómez-Caravaca, A.M. et al. Determination of the major phenolic compounds in pomegranate juices by HPLC−DAD−ESI-MS. *J. Agr. Food Chem.* 2013, 61, 5328-5337.

[12] Saad, H. et al. Characterization of pomegranate peels tannin extractives. *Ind. Crop. Prod.* 2012, 40, 239-246.

[13] Abdel-Hady, N. M. *J. Appl. Sci. Res.* 2013, 9, 4823-4830.

[14] Qu, W. et al. *Food Chem.* 2012, 132, 1585-1591. Quantitative determination of major polyphenol constituents in pomegranate products.

[15] Qu, W., Pan, Z., Ma, H. *J. Food. Eng*. 2010, 99, 16-23. Extraction modeling and activities of antioxidants from pomegranate marc.
[16] Zhao, X. et al. Characterization and evaluation of major anthocyanins in pomegranate (Punica granatum L.) peel of different cultivars and their development phases. *Eur. Food Res. Technol.* 2013, 236, 109-117.
[17] Zhang, Y., Wang, D., Lee, R-P., Henning, S.M., Heber, D. Absence of pomegranate ellagitannins in the majority of commercial pomegranate extracts: Implications for standardization and quality control. *J. Agric. Food Chem.* 2009, 57, 7395–7400.
[18] Richter, B. E. et al. *Anal. Chem.* 1996, 68, 1033-1039.
[19] Smith, R. E. et al. Noni composition and health benefits, in *Fruit Juices: Types, Nutritional Composition and Health Benefits*, Elder, K.E., ed. Nova Sciences, Hauppauge, NY, 2014.
[20] Nuncio-Jáuregui, N. et al. *Sci. Hort.* 2014, 165, 181-189.
[21] Zhao, X. et al. Characterization and evaluation of major anthocyanins in pomegranate (Punica granatum L.) peel of different cultivars and their development phases. *Eur. Food Res. Technol.* 2013, 236, 109-117.
[22] Fawole, O. A., Opara, U. L. *Sci. Hort.* 2013, 159, 152-161.
Developmental changes in maturity indices of pomegranate fruit: A descriptive review.
[23] Al-Maiman, S. A., Ahmad, D. Changes in physical and chemical properties during pomegranate (*Punica granatum* L.) fruit maturation. *Food Chem.* 2002, 76, 437-441.
[24] Fawole, O. A., Opara, U. L. Effects of maturity status on biochemical content, polyphenol composition and antioxidant capacity of pomegranate fruit arils (cv. 'Bhagwa'). *S. Afr. J. Botany* 2013, 85, 23-31.
[25] Zhang, Y. et al. International multidimensional authenticity specification (IMAS) algorithm for detection of commercial pomegranate juice adulteration. *J. Agric. Food Chem.* 2009, 57, 2550-2557.
[26] Nuncio-Jáuregui, N., Calín-Sánchez, A., Hernández, F., Carbonell-Barrachina, A. A. *J. Sci. Food Agric.* 2014, 94, 646–655. Pomegranate juice adulteration by addition of grape or peach juices.
[27] Kulkarni, A. P., Aradhya, S. M. Chemical changes and antioxidant activity in pomegranate arils during fruit development. *Food Chem.* 2005, 93, 319-324.
[28] Lee, J.; Durst, R. W.; Wrolstad, R. *EJ. AOAC Int.* 2005, 88, 1269-1278.
[29] Mena, P. et al. *Food Chem.* 2012, 133, 108-115.
[30] Mayuoni-Kirshinbaum, L., Daus, A., Porat, R. Changes in sensory quality and aroma volatile composition during prolonged storage of 'Wonderful' pomegranate fruit. *Food Sci. Technol.* 2013, 48, 1569-1578.
[31] Andreu-Sevilla, A. J., Mena, P., Marti, N., Viguera, C. G., Carbonell-Barrachina, Á, A. Volatile composition and descriptive sensory analysis of pomegranate juice and wine. Food Res. Int. 2013; 54, 246-254.
[32] Cama, M., Hisil, Y., Durmaz, G. Classification of eight pomegranate juices based on antioxidant capacity measured by four methods. *Food Chem.* 2009, 112, 721-726.
[33] Abdel-Hady, N.M. Quantitative Diversity of Phenolic Content in Peels of Some Selected Egyptian Pomegranate Cultivars Correlated to Antioxidant and Anticancer Effects. *J. Appl. Sci. Res.* 2013, 9, 4823-4830.
[34] Ullah, N. et al. *Middle East J. Sci. Res.* 2012, 11, 396-401.
[35] Hasnaoui, N., Wathelet, B., Jiménez-Araujo, A. *Food Chem.* 2014.

[36] Kaufman, M., Wiesman, Z. *J. Agric. Food Chem.* 2007, 55, 10405-10413.

[37] Fadavi, A., Barzegar, M., Azizi, M. H. Determination of fatty acids and total lipid content in oilseed of 25 pomegranates varieties grown in Iran. *J. Food Comp. Anal.* 2006, 19, 676-680.

[38] P Melgarejo, P., Artés, F. *J. Sci. Food Agric.* 2000, 80, 1452-1454.

[39] Sassano, G. et al. Analysis of pomegranate seed oil for the presence of jacaric acid. *J. Sci. Food Agric.* 2009, 89, 1046-1052.

[40] Zhang, L., Gao, Y., Zhang, Y., Liu J., Yu, J. Changes in bioactive compounds and antioxidant activities in pomegranate leaves. *Sci. Hort.* 2012, 123, 543-546.

[41] Bekir, J., Mars, M., Souchard, J.P., Bouajila, *J. Food Chem. Toxicol.* 2013, 55, 470-475.

[42] Gómez-Caravaca, A.M. et al. Determination of the major phenolic compounds in pomegranate juices by HPLC−DAD−ESI-MS. *J. Agr. Food Chem.* 2013, 61, 5328-5337.

[43] Hajimahmoodi, M. et al. *Am. J. Plant Sci.* 2013, 4, 1815-1830.

[44] Hajimahmoodi, M. et al. Total Phenolic, Flavonoids, Tannin Content and Antioxidant Power of Some Iranian Pomegranate Flower Cultivars (*Punica granatum* L.). *Amer. J. Plant Sci.* 2013, 4, 1815-1820.

[45] Fu, Q. et al. *Food. Bioprod. Process.* Published on-line in, 2013.

[46] Capriotti, A.L. et al. *Anal. Bioanal. Chem.* 2013, 405, 9301-9309.

*Chapter 5*

# Health Effects

## Introduction

Many ancient cultures believed in the health-promoting effects of all parts of the pomegranate tree, including the fruit, bark, flowers, roots, and leaves [1]. These are due to the various bioactive phytochemicals in pomegranates [1, 2]. The concentrations and amounts of these compounds in the pomegranate tree change during its development and fruit maturation. They also depend on the environmental conditions and which cultivars are being grown [1].

Egyptians used pomegranates to treat infections [3]. Extracts from the peels of the fruit and bark of the tree have been used thousands of years in the Ayurvedic system of medicine for diarrhea and dysentery [4]. Pomegranates are also used as an antiparasitic agent and to treat ulcers [5]. They are also used in the Unani system of medicine in the Middle East and India to treat diabetes [5]. In modern medical research, there is evidence supporting its possible use in helping to prevent and/or treat cancer, cardiovascular disease, diabetes, dental conditions, erectile dysfunction, infant brain ischemia, Alzheimer's disease, male infertility, arthritis, and obesity as well as protect from ultraviolet (UV) radiation [5]. Many of the health effects may be due to synergistic interactions between the various bioactive compounds in pomegranates and products made from them [5]. Moreover, there are dietary supplements that contain extracts of pomegranates. Some of the ingredients (such as ellagic acid) are highly enriched, while others are not present. This can affect the health properties and the attitudes of the consumers. Some believe that no man-made extract can possibly be as effective as the whole food, so they might be reluctant to buy or consume extracts.

## Safety and Toxicity Studies

For new chemical entities, the first step in assessing its health properties is to study its potential toxicity. This is not necessary for the pomegranate fruit, seed oil, juice or extracts prepared from them, because they are generally regarded as safe (GRAS) in the USA and other countries [6]. However, the root and stem barks have some toxicity, due to the presence of alkaloids [7].

Even though extracts are GRAS, there was still a toxicity study done on rats, in which high doses (up to 6% of their feed) of the ellagitannin called punicalagin were given for 37 days and were non-toxic [7]. As in any properly conducted feeding study, the authors first did a preliminary study in which they established the acceptability of the extract by the animals. That is, a test article or new chemical entity can affect a rodent's health if it tastes so good (or bad) that it causes them to increase (or decrease) their consumption of the rodent feed. This can affect the amount of weight that they gain (or lose) during the study.

So, the authors extracted one kg of dried peels (husks) of the 'Mollar de Albatera' variety with 2 L of distilled water after being stirred vigorously and left to stand at room temperature for two hrs [7]. "This was repeated several times. The combined extracts were freeze-dried and a portion of them was added to the rodent feed to get a ratio of 1:4 pomegranate extract to rodent feed. The concentration of punicalagin was 6% of the mixture based on LC-MS-MS analysis. This compares to a concentration of 2 g/L in juice. The rats consumed an increasing amount of punicalagin each day: 0.39, 0.54, 1.12, 1.3, and 1.18 g/day, during the first, second, third, fourth, and fifth weeks, respectively. The average was 0.9 g/day, which corresponds to a 70 kg human consuming about 194 L/day" [7].

Even though this may seem to be a ridiculously high amount, it is how rodent toxicology studies are done. The idea is that there might be a very small portion of the human population that may be highly susceptible to the test article or new chemical entity. It is not feasible to test chemicals on thousands or millions of rodents, so toxicologists give a few of them doses that are much larger than what could be consumed by anyone. Supposedly, this gives the toxicologist an idea about the potential toxicity that might affect a small portion of people [8]. Still, such studies can identify metabolites and be used to see if the test article can reach the target organs. So, this study [7] found gallic acid and two ellagic acid derivatives in the liver and kidneys: 3,8-dihydroxy-6*H*-dibenzo[*b,d*]pyran-6-one glucuronide and 3,8,10-dihydroxy-6*H*-dibenzo[*b,d*]pyran-6-one glucuronide.

After establishing the safety in rodents, the next step is to test the safety in healthy people [8]. In many ways, this has already been done, since pomegranates and juice made from them have been consumed for several millenia and are GRAS. However, this was reinforced by some studies done on healthy human subjects [9-11]. In one study, one person was given 180 mL of juice that contained 25 mg of ellagic acid (EA) and 318 mg ellagitannins [9]. EA was detected in the blood plasma. Its maximum concentration was 31.9 ng/mL one hour after ingestion, but decreased after that until it was below the detection limit after 4 hours [9].

In a subsequent study, the authors gave 16 healthy people 8 ounces (about 227 g) of pomegranate juice (var. 'Wonderful'), 8 ounces of a pomegranate polyphenol liquid extract (POMx1) and 1000 mg of a pomegranate polyphenol powder extract (POMxp) with a one week washout period in between [10]. "There was no statistically significant difference in the areas under the curve (AUC, see Chapter 1) for the three treatments. The AUCs were about 0.11 μmol-hr/L. The average times required to reach the maximum concentration were 0.65, 0.94 and 2.58 hrs for the juice, POMx1 and POMxp, respectively. A metabolite of ellagic acid, urolithin-A glucuronide reached about 1000 ng/mL in each treatment" [10].

A different group measured the absorption, metabolism and antioxidant effects of 800 mg of a standardized pomegranate extract containing 330 mg of the major ellagitannins and 21.6 mg of EA, as determined by HPLC [11]. "Thirteen healthy men and women who were neither pregnant nor lactating (BMI = 32.6 ± 0.98, age = 37.6 ± 3.6) were asked to abstain from eating foods that contain polyphenols, such as tea, wines and berry fruits for three days. They were also were asked not to consume large amounts of alcohol and antioxidant supplements, not to exercise excessively, and to sleep at least 6-8 hrs during the night before the study day.

The washout period ended with an overnight fast (at least 8 hrs) before the study day. They found that ellagic acid was bioavailable, with a maximum concentration of 33 ng/mL after 0.5 hrs. Urolithin A, urolithin B, hydroxyl-urolithin A, urolithin A-glucuronide, and dimethyl ellagic acid-glucuronide were identified as metabolites in the blood plasma. The antioxidant capacity increased by 32% after 0.5 hrs, but the generation of reactive oxygen species was not affected, nor was a marker of inflammation (interleukin-6)" [11].

It is also important to look at anthocyanins, which also have important health effects and are found in many fruits. One group found that a small portion (6%) of them were bioavailable, as were a larger amount of their metabolites [12]. They were found in the urine after drinking 250 mL of blueberry juice. Even though this study did not test pomegranate juice, it might be reasonable to think that they are also bioavailable.

## Antioxidant and Anti-inflammatory Properties

Still, it is well known that many dietary antioxidants have low bioavailabilities. Moreover, urate and plasma proteins contribute much more to the antioxidant capacity of blood plasma than do dietary antioxidants [13]. Also, giving large doses of dietary antioxidant supplements to human subjects has seldom had any preventative or therapeutic effects [14]. This is called the antioxidant paradox [14]. However, it is possible that dietary phenolic compounds can be quite healthy and prevent oxidative damage by forming chelates or slightly dissociated complexes with iron ($Fe^{2+}$) and copper ($Cu^{2+}$), preventing them from making the very dangerous hydroxyl radicals in the Fenton reaction [15]. So, all of the *in vitro* tests for antioxidant activities of pomegranates, juices and extracts could be biologically and medically irrelevant. As mentioned by others [14, 15], consuming dietary antioxidants may not be as helpful as not eating meat and not taking dietary supplements that contain iron or copper, unless a medical condition such as anemia dictates their consumption. In fact, iron has been removed from dietary supplements for men over 50 for quite some time [16]. So, as potential health effects are discussed in the rest of this chapter, remember that complex problems (like diseases) usually have multiple causes [8]. There are many things that can be done to prevent and treat diseases. Consuming pomegranates, their juice or products made from them can be a part of an overall health care program, but this is unlikely to overcome bad health habits and other factors. Still, low-grade, smoldering inflammation is the root cause of many diseases, including diabetes, stroke, cardiovascular diseases, many types of cancer, autoimmune diseases and neurodegenerative diseases [8]. Pomegranates may help prevent and/or alleviate smoldering inflammation.

So, there have been many articles written about the *in vitro* antioxidant properties of pomegranates and its juice [17-24]. Their contents will be described next.

The *in vitro* antioxidant effects of punicalagin, ellagic acid and a total pomegranate tannin extract were analyzed for their abilities to inhibit the peroxidation of lipids induced by $Fe^{2+}$ and to prevent the oxidative damage caused by a water-soluble free radical called 2,2'-azinobis(3-thylbenzothiazline-6-sulfonic acid)-diammonium salt, or ABTS [17]. That is, oxidative damage can occur in cell membranes and the lipid-containing membranes of internal organelles. It can also occur in the aqueous cytosol, as measured by ABTS. In the test for lipid peroxidation, artificial membranes called liposomes [8] are used. The ability of a test compound or mixture of compounds is compared to the standard antioxidants, *tert*-butylhydroquinone (TBHQ), butylated hydroxyanisole (BHA) and butylated hydroxytoluene (BHT). It was found that pomegranate juice was able to prevent lipid peroxidation better than the standardized total pomegranate tannin extract, punicalagin and EA [17]. "Similarly, pomegranate juice was the most active in the aqueous assay, which uses a compound called Trolox as a standard and positive control. The Trolox equivalent antioxidant capacities (TEACs) of the juice, standardized total pomegranate tannin extract, punicalagin and EA were 25591, 100, 90 and 40 μmol Trolox equivalents, respectfully. That is, the juice was a more potent antioxidant than were the separated and purified compounds present in it, suggesting synergistic interactions" [17].

At the same time this article reported that the juice, standardized total pomegranate tannin extract, punicalagin and EA all exhibited antiproliferative and apoptotic effects against human colon and prostate tumor cell lines [17]. As mentioned in Chapter 1, one of the important properties of living organisms and cells is autopoiesis, or self-making. That is, most of the cell membranes and cells in our bodies are continuously being broken down and re-made [8]. Cells are supposed to die at appropriate times in a process called programmed cell death or apoptosis. When cells do not undergo apoptosis, they can become cancer cells in a tumor. So, many anticancer drugs work by activating apoptosis [8]. In addition, some of the anticancer effects of pomegranates and its juice may be due to their antiproliferative and pro-apoptotic properties.

Another article described the phenolic content of something they called pomegranate meal [18]. "It was the material that remained after cold pressing. They extracted the de-fatted meal three times with methanol/water (7:3, v/v) in an ultrasound bath at room temperature. The supernatants were analyzed for free phenolic acids and soluble phenolic acid esters, while the residue was analyzed for insoluble, bound phenolic acids. The predominant phenolic acid was gallic acid, but protocatechuic, *p*-hydroxybenzoic and vanillic acids were also found in the free form by HPLC. They were also esters of these and some insoluble bound forms. There was no free caffeic or syringic acid, but there were esters of them. The concentrations of free, esterified and insoluble bound gallic acid were 1179, 581 and 707 mg/kg dry weight, respectively. The total free, esterified and bound phenolic acids were 1210, 663 and 729 mg/kg dry weight, respectively" [18]. Moreover, caffeic acid is not just a good antioxidant but it also chelates metals, including $Fe^{2+}$ and $Cu^{2+}$. Gallic acid can also chelate metals and has a binding constant (4.78 $M^{-1}$) that is 59% of that of caffeic acid (8.12 $M^{-1}$) [25]. So, pomegranate meal and even pomegranates and juice may quite effective in preventing smoldering inflammation, diabetes, cancer, stroke, cardiovascular diseases and autoimmune diseases by preventing poorly liganded $Fe^{2+}$ and $Cu^{+2}$ from undergoing the Fenton reaction (Reaction 1) that produces the highly reactive and damaging hydroxyl radical (OH•) [8, 15].

$$Fe^{2+} + H_2O_2 \rightarrow Fe^{3+} + OH^- + OH\bullet$$

Reaction 1. Fenton reaction.

Pomegranate juices from Turkey were also found to contain phenolic acids, as well as other organic acids and sugar [19]. "The total concentrations of total phenolics were measured by a colorimetric assay in which they react with a reagent to produce a dark purple color. This is a widely used method that requires using a phenolic compound as a standard – usually gallic acid. So, results are expressed as mg of gallic acid equivalents (GAE) per L of juice (or per g of dry samples). The total phenolics ranged from 144 to 10086 mg GAE/L for seven different juice samples. They also measured the *in vitro* antioxidant activities of the juices. This was done measuring the ability of the juice to prevent oxidation of the DPPH radical (see Chapter 4). The antioxidant activities based on the DPPH assay ranged from 10.4 to 67.5% inhibition of DPPH oxidation. The sample that had the highest concentration of total phenolics and the highest DPPH antioxidant activity. They also measured the ferric ion reducing antioxidant capacity or FRAP" [19]. When $Fe^{3+}$ is reduced to $Fe^{2+}$, the $Fe^{2+}$ binds to a colorimetric reagent and forms a colored complex that absorbs at 593 nm [26]. A standard curve was prepared by using different concentrations of $Fe^{2+}$ [19]. "The FRAP antioxidant capacity ranged from 10.4 to 67.5 mmol/L of $Fe^{2+}$. The concentrations of citric acid, glucose and fructose ranged from 3.93 – 13.1, 39.8 – 65.4 and 45.5 – 93.6 mg/L, respectively. Finally, one sample had much more malic acid than it should, suggesting that it had been adulterated with apple juice" [19].

The antioxidant activity of Californian pomegranate juice was found to depend on how the juice was prepared [21]. "It was higher in commercial juices prepared from whole pomegranates compared to juices made from just the arils of the same 'Wonderful' cultivar. This was due (at least partly) to the fact that the antioxidant punicalagin was present in the commercial juices (1500 – 1900 mg/L), but not juices prepared from arils. They used four different methods to measure antioxidant activity. Two of them, DPPH and FRAP have already been discussed. The other two were based on ABTS (2,2'-azino-bis(3-ethylbenzothiazoline-6-sulphonic acid)) and DMPD (N,N-dimethyl-p-phenylenediamine). The ABTS assay uses $H_2O_2$ and horseradish peroxidase to produce the colored ABTS radical ($ABTS^{\cdot+}$) that absorbs light at 414 nm. When antioxidants are added, they prevent the formation of $ABTS^{\cdot+}$. In the DMPD assay, $Fe^{3+}$ is added to *p*-phenylene diamine to produce the colored free radical $DMPD^{\cdot+}$ that absorbs at 505 nm. Antioxidants prevent this from happening. The antioxidant activities of the pomegranate juices were three times higher than those of green tea and red wine. They were due to anthocyanins (delphinidin 3,5-duglucoside, cyanidin 3,5-diglucoside, delphinidin 3-glucoside, cyanidin 3-glucoside and pelargonidin 3-glucoside), gallyl type tannins (punicalagins B and D as well as others), ellagic acid derivatives (ellagic acid and ellagic acid glucoside as well as others) and other hydrolyzable tannins (galloyl glucose and others). The concentrations of total phenolics were 2117 and 2566 in juices made from fresh arils and commercial juice, respectively" [21].

Whole pomegranates and pomegranate peels from Taiwan were shown to have *in vitro* and *in vivo* anti-inflammatory effects [20]. This study was conducted because pomegranates are widely used as an antipyretic analgesic in Chinese culture [20]. However, there were some problems with this study. First, they said that test solutions of pomegranate were prepared by dissolving pomegranate in 10% DMSO (dimethyl sulfoxide). This was then diluted into cell culture growth medium and added to a mouse (murine) macrophage cell line. The first major problem is that the authors did not say if they used whole pomegranates or just the peels, which they said were used to isolate polyphenols. Second, neither the whole pomegranate nor any its anatomical parts of it are soluble in 10% DMSO. Even though many of the isolated bioactive compounds are soluble in 10% DMSO (with 90% water), they may or may not be extracted by it, especially if one just shakes pomegranates with the 10% DMSO. The amount extracted will depend on how long the sample was shaken or sonicated. Moreover, once the 10% DMSO is diluted into aqueous growth medium, many of the relatively hydrophobic bioactive compounds that are dissolved in it will precipitate out of solution and not enter the cells. So, this study was not done using good laboratory practices (GLP), which require that a dose preparation and analysis study be done before giving a dose to cell cultures or rodents. For a GLP study, a dose must be proven to be homogeneous. If solids settle out of the dose (as it would with a mixture of pomegranates, DMSO and aqueous growth medium), it can't be used. So, the actual doses of bioactive compounds given to the cell culture were unknown and probably quite variable in the study mentioned [20].

The second problem with the study [20] is that the authors described doing assays for total phenols and flavonols on a pomegranate extract that was not described. They did say in a separate section that they isolated polyphenols from peels using 70% aqueous acetone, but the reader is left to guess whether this solvent was used to prepare the pomegranate extract or whether whole pomegranates or just peels were used. The third major problem is that they used a relatively ineffective method to extract polyphenols. They simply mixed 600 g of dried peels with 3 L of 70% acetone, homogenized the mixture and filtered it. This was repeated three more times, so about 12 L of acetone, a dangerous greenhouse gas, were required. They could have extracted mush more material and used much less solvent by using dry methanol at 100 $^{o}$C and 10 MPascal pressure in a sealed container [27, 28]. Still, they were able to extract some of the bioactive compounds that are antioxidants [20].

They reported finding 471 mg GAE/g of total phenols in the 70% acetone extract [20], but it is quite possible that they actually meant that this was the concentration in the actual pomegranates (or peels) and not the extract. That is, if the 471 mg GAE was in one gram of the 12 L (about 12 kg) of extract, it would mean that there were 5652000 mg of GAE in 600 g of peels, or 9420 mg GAE per gram of pomegranates (or peels). This is much more than 471 mg GAE per gram. They also reported finding 257 mg of flavonols (as catechin equivalents) per g of 70% acetone extract [20], but it is probably the concentration in whole pomegranates or peels.

They also reported that some of the bioactive compounds had *in vitro* and *in vivo* anti-inflammatory effects [20]. That is, they prevented *in vitro* inflammation in cell cultures of mouse macrophages that were insulted with lipopolysaccharides (LPS), which caused inflammation in control cells that did not have added pomegranate extracts [20]. That is, LPS induces the production of pro-inflammatory cytokines, such as prostagalndins and nitric oxide (NO). Moreover, inflamed macrophages can lead to atherosclerosis and cardiovascular diseases, which will be discussed further in the Appendix. They also found that some of the bioactive compounds prevented edema on the paws of mice that were exposed to the pyrogen carrageenan [20]. The hydrolyzable tannins, punicalagin, punicalin, strictinin A and granatin B were isolated and shown to be anti-inflammatory [20]. Still, it is quite possible that pomegranates and their peels might be much more active and have much higher concentrations of these and other bioactive antioxidants. They might also work synergistically. So, the basic conclusion of this study is valid. Pomegranates do have important anti-inflammatory properties.

A commercial pomegranate extract (POMx) was found to lower lipid peroxidation in obese adults with type-2 diabetes but without complications or requiring insulin, but not in healthy people [22]. The diabetics took two capsules of POMx containing 753 mg of polyphenols per capsule daily for four weeks. Their blood pressure, body weight and concentrations of glucose and lipids in the blood were not affected significantly. However, the concentrations of two markers of lipid peroxidation (malondialdehyde, MDA, and hydroxynonenal, HNE) were reduced [22]. Two POMx capsules weighed 2.1 g and contained 1.8 g of carbohydrates, 5.86 mg of fat, 72.5 mg of protein, 113 mg of ash, 116 mg of moisture 372 mg ellagic acid and 1505 mg of total phenolics [22].

## REDUCING WASTE AND PREVENTING MALNUTRITION

Another health benefit that few people think of is reducing agroindustrial waste and preventing malnutrition. By reducing the production of agroindustrial waste, fewer greenhouse gases like carbon dioxide and methane will be produced as less waste is transported to landfills and left to decay. So, as peels and seeds are used for other purposes, society will benefit. For example, pomegranate peels can be used to prepare phenolic compounds that can be added as a nutritional ingredient to foods [23]. One group used peels that were a by-product of the Turkish juice industry [23]. "They were dried naturally at room temperature for 5 days on trays that were kept out of the sunlight. Phenolics were extracted from 200 g of dried peels using 1 L of boiling water for 5 min, then cooled to room temperature. The concentration of total phenolics in the extract was about 29.5 g/L. This was adjusted to about 20.0 g/L by adding water. Portions of this were mixed with maltodextrins, which serve as a coating material for the core material (phenolics). Microcapsules were made by spray drying. Different ratios of phenolics to maltodextrin were tested. The optimum ratio was found to be 1:1. The microcapsules were stored at 4 °C for three months to determine their stability. For quality assurance, the color and total phenolic content were measured. Moreover, individual phenolic compounds were separated by ultra performance liquid chromatography (UPLC). Then, the microcapsules were added to a low fat Turkish ice cream formulation until it contained either 0.5 or 1.0% encapsulated phenolics. This was evaluated by sensory analysis to see if the ice cream was acceptable to consumers" [23].

The main phenolics in the extracts and microcapsules were punicalagins (80.3%) and ellagic acids (19.7%) [23]. "The ice cream containing 1.0% microcapsules had an antioxidant activity of 133 g per g. The ice cream mixes were pasteurized at 80 °C for 10 min, but this did not significantly decrease the concentration of total phenolics. The phenolic compounds in it also inhibited the enzyme α-glucosidase, which catalyzes the hydrolysis of starch and disaccharides to make glucose. So, these two classes of carbohydrates will be digested slower, producing less glucose, which can help prevent type-2 diabetes. Also, over 75% of the taste testers liked the taste of the ice cream. Also, they did not feel any astringency, or dry puckering feeling in the mouth that can occur when consuming tannins" [23]. So, if a person can keep from eating too much of this ice cream (and consuming too many calories), this tasty treat could be a healthy addition to a diabetic diet.

Extracts of pomegranate seeds have also been added to meat to decrease the formation of cancer-causing heterocyclic aromatic amines when the meat is cooked [24]. "They are produced at ng/g levels by the Maillard reaction between creatine, sugars and amino acids when meat and fish are cooked at 150 °C or higher. Commercially available seed extract from Turkey was added to the ground meat at 0.5% (w/w). The concentration of total phenolics was 353 GAE/g. Chicken and beef were cooked by the most commonly used methods in Turkey: pan cooking, oven roasting, charcoal barbecue, and deep-fat frying. For deep-fat frying, fresh sunflower oil was used. When the temperature of oil increased to 150 °C, the meatballs were fried for 5 min. The concentrations of total phenolics were 201, 231, 238 and 330 mg GAE/g when beef was cooked by pan cooking, deep fat frying, charcoal barbecue and oven roasting, respectively. The concentrations of total phenolics were 231, 245, 326 and 389 mg GAE/g when chicken was cooked by deep fat frying, pan cooking, oven roasting and charcoal barbecue, respectively. The antioxidant activity of the beef and chicken meatballs depended on the cooking method. Deep-fat frying of beef produced meatballs with the lowest DPPH radical scavenging activity, while oven roasted meatballs had the highest. In all cooking methods, the radical scavenging activity increased when pomegranate seed extract was added. It increased from 16.95 to 33.10% for deep-fat fried beef and from 23.99 to 51.50% for oven roasted. For chicken meatballs, the radical scavenging activity was lowest when cooked in a pan and highest when charcoal barbecued. It increased from 24.95% to 48.52% when cooked in a pan and from 36.83% to 58.26% when charcoal barbecued" [24].

Pomegranate seed extract also decreased the formation of the cancer-causing heterocyclic aromatic amines 2-amino-3-methylimidazo[4,5-f]quinoline (IQ), 2-amino-3,8-dimethylimidazo [4,5-f]quinoxaline (MeIQx), 2-amino-1-methyl-6-phenylimidazo [4,5-b]pyridine (PhIP), and 9H-pyrido[3,4-b]indole (norharman) [24]. "The seed extract reduced the formation of IQ by 38% and 45% in charcoal-barbecued and deep-fat fried beef meatballs. It reduced the formation of IQ by 46% in deep-fat fried chicken. It also reduced the formation of MeIQx by 57% in beef meatballs when they were charcoal barbecued and by 49% in deep-fat fried chicken. It reduced the formation of PhIP by 62% and 68% in beef meatballs cooked in a pan or barbecued with charcoal, respectively. It reduced the formation of PhIP by about 70% in chicken cooked by oven roasting, pan cooking, and charcoal barbecuing. It reduced the formation of norharman by 24% in charcoal-barbecued beef and 50% in chicken cooked by all methods tested" [24]. It should be noted that even better results might be obtainable if pomegranate seeds were to be extracted with pressurized water or especially pressurized ethanol. Of course, it would take much less electricity to evaporate or distill off the ethanol than water.

As mentioned previously, a commercial pomegranate extract (POMx) was found to lower lipid peroxidation in obese adults with type-2 diabetes but without complications or requiring insulin, but not in healthy people [22]. One of the most dangerous products of lipid peroxidation is 4-hydroxynonenal, abbreviated as HNE [29]. It is an α,β-unsaturated aldehyde that is produced by the radical-mediated peroxidation of ω-6 polyunsaturated fats. It is a bioactive molecule that can help cause several diseases if present at too high of a concentration. Most of its biological activities are due to it reacting with biomolecules (like proteins, DNA and phospholipids). However, when present at low concentrations, HNE is an endogenous signaling molecule [22]. This illustrates an important point – toxicity depends on the dose. In other words, all substances (including fresh air and fresh water) are toxic if given at too high of a dose to the most susceptible parts of the body. Two other ways to say this is that the dose is the poison and the difference between pharmacology and toxicology is the dose. Still, the article cited [29] makes the point that polyphenols like those present in pomegranates can reduce the production of HNE by not just their antioxidant properties, but also their ability to bind or chelate metal ions [29]. This re-emphasizes the point made earlier, that poorly liganded $Fe^{2+}$ and $Cu^{+2}$ can undergo the Fenton reaction that produces the highly reactive and damaging hydroxyl radical [8, 15].

## ANTIOXIDANTS IN POMEGRANATE JUICE

Still, there can be other important mechanisms by which antioxidants in pomegranates can have important health effects. One important example is the antioxidant punicalagin, which is present in pomegranate juice [30]. It may be able to help prevent preeclampsia and intrauterine growth restriction (IUGR) [30]. Both conditions have similar placental histopathology [31, 32]. They can cause improper remodeling of the maternal spiral arteries and damage the chorioallantoic placenta after oxidative stress [30 – 32]. This can lead to progressive damage in villi and cause placental dysfunction, which can harm or even kill the mother and her baby [30].

Also, it was previously demonstrated that pomegranate juice reduces oxidative stress in human villous trophoblasts and limits stimulus-induced cell death of human trophoblasts in explants and in primary cultures [33]. This was supported by another study that showed that an extract of the fruit could prevent oxidative stress in the fetus caused by the anticancer agent, adriamycin. [34]. These antioxidant effects may be due to the antioxidant potential of the phenolic compounds and/or their abilities to chelate $Fe^{2+}$ and $Cu^{2+}$. This could be important in helping to prevent cardiotoxicity, which can occur in cancer patients (especially children) who are being treated with adriamycin (also known as doxorubicin). This is because poorly liganded (or chelated) $Cu^{2+}$ can increase the toxicity of adriamycin [35]. However, punicalagin can also downregulate or decrease the level of an important regulatory protein in the body called p53 [30]. This is one of the most important proteins in human cells and is a hub in the cellular network and web of human life, as discussed further in the Appendix. It is often called the guardian of the genome. Its role is to ensure that any damage to DNA is repaired before it allows cells to progress in the mitotic cell cycle as two genetically identical cells are produced from a parent cell. It can also activate programmed cell death, or apoptosis.

So, researches showed that punicalagin and pomegranate juice decrease oxidative stress and apoptosis in cultured cells called syncytiotrophoblasts that were exposed to hypoxia (low $O_2$) [30]. These are cells in the epithelium that covers the embryonic placental villi, which enter the wall of the uterus to enable nutrients to circulate between the mother and the growing fetus [36]. Placental apoptosis increases during preeclampsia and intrauterine growth restriction [36]. In healthy pregnancies, the ratio of cytotrophoblasts and syncytiotrophoblasts is maintained in a steady state, but it is altered in pathological conditions [36]. Even though some apoptosis is needed to control the number of cells and size of tissues [36], excess placental apoptosis of trophoblasts occurs in preeclampsia and intrauterine growth restriction [37, 38]. Oxidative stress can cause placental dysfunction and problems in pregnancy, and p53 is involved in apoptotic stress caused by oxidative stress [30]. So, it is quite significant that punicalagin and pomegranate juice can limit oxidative injury by helping to regulate p53 activity [30]. Moreover, p53 causes apoptosis by activating enzymes called caspases, which catalyze the hydrolysis or breakdown of cellular proteins. So, it is important to note that punicalagin the expression and subsequent activity of caspases [30]. The p53 protein also affects the activities of two other proteins, p21 and MDM2, both of which were affected by punicalagin [30]. Finally, there is a protein called heat inducible factor1α or HIF1α that is induced by hypoxia, interacts with MDM2 and alters p53 activity and apoptosis [30]. Punicalagin reduced the expression of HIF1α in syncytiotrophoblasts [30].

## Osteoarthritis

A pomegranate extract was also shown to inhibit the expression of an important set of enzymes called matrix metalloproteinases (MMPs) [39]. These enzymes catalyze the hydrolytic breakdown of proteins in many physiological processes [40] including type-2 collagen from the cartilage [41]. "Together with inflammation, this breakdown can cause osteoarthritis, the most common form of joint disorder. It is associated with age-related changes in subchondral bones and progressive erosion of articular cartilage. This can lead to the loss of joint function and osteoarthritis. There is an imbalance between the formation and breakdown of cartilage in osteoarthritis. When inflammation occurs, the proinflammatory cytokine interleukin IL-1β is produced in the affected joints. It upregulates or increases the activity of MMPs, leading to the irreversible breakdown of cartilage and subsequent release of proteoglycan from the matrix. A methanolic extract of pomegranate powder at doses of 6.25 – 25 mg/L inhibited the expression of MMPs caused by IL-1β. This inhibited the degradation of cartilage, so it could be useful in maintaining the integrity and function of joints" [41]. However, the procedure used to prepare the methanolic extract was not described. So, it is quite possible that this paper [41] under-estimated the ability of pomegranates or a methanolic extract of them to help prevent or treat osteoarthritis.

## Obesity and Metabolic Syndrome

Osteoarthritis, chronic smoldering inflammation, diabetes, kidney disease, cardiovascular disease, cancer and neurodegenerative diseases all share a common risk factor – obesity or metabolic syndrome [8]. The possible role of pomegranates in preventing obesity was reviewed recently [42]. They cited numerous studies, starting with one in which mice on a high fat diet were fed pomegranate seed oil (1 g per 100 g body weight, or 1 g/100 g BW) [43]. The total weight of their fat decreased compared to controls [43]. In another study, a dose of 1 g/100 g DW of a conjugated linolenic acid called catalpic acid was found to be an important ingredient in seed oil that improved (lowered) fasting glucose and abdominal white adipose tissue in mice [44]. This is important, because white adipose tissue is now known to be part of the endocrine system since it produces adipose hormones (adipokines) that affect appetite and satiety [8]. Catalpic acid also increased the concentration of high-density lipoproteins (HDL) in blood, while decreasing triglycerides [44]. Pomegranate leaves, which are used to make some types of green tea in China, were found to decrease the body weight, abdominal fat and adipose fat pad weight [45]. They also decreased cholesterol, triglycerides and the ratio of total cholesterol to HDL [45]. Pomegranate flowers, which are used as a traditional antidiabetic diabetic medicine, corrected faulty cardiac lipid metabolism in diabetic fat rats by activating peroxisome proliferator activated receptor-γ (PPAR- γ), which is also the therapeutic target for prescription drugs like pioglitazone, which is used to treat diabetes [8]. Another group found that extracts from pomegranate flowers lowered the concentration of lipids and glucose in the blood serum in mice [46]. Reportedly, 10 mg of flowers were ground into a powder and dissolved in 1 mL of water [46], but only a small portion of the flowers can dissolve in water. So, it is possible that the effects on lipid metabolism would be much higher if the whole flowers were consumed, or if pressurized water or ethanol were used in the extraction. Also, two important components of pomegranate flowers, oleanolic acid and ursolic aicd have antidiabetic properties [47, 48]. Another study showed that oleanolic acid (10 mg/kg BW) decreased the blood glucose concentration as well as the body weight and visceral abdominal fat in obese mice [49]. Also, phenolic compounds in flowers, gallic acid, corrected hyperlipidemia and fatty liver in mice on a high fat diet [50]. Flower extract decreased the concentration of triglycerides in rats that had diabetic fatty liver disease [51]. Another study found that punicalagin in pomegranate juice lowered food consumption and body weight in female rats [52].

Pomegranate juice may also be effective in people [53]. Twenty-two diabetic patients were given concentrated juice (40 g/day for 8 weeks). Their blood levels of cholesterol, LDL cholesterol/HDL cholesterol, and total cholesterol/HDL cholesterol decreased [53, 54]. Also, consuming pomegranates and products made from them caused ellagitannins and punicalagins to appear in the large intestines. They interacted with complex gut bacteria and inhibited the growth of pathogenic *Clostridia*, *pseudomonas aeruginosa*, and *Staphylococcus aureus* [42, 54]. Multi-resistant *S. aureus* or MRSA is an emerging threat to human health, since it is resistant to most antibiotics [8]. The study [54] also showed that probiotic lactobacilli and most bifidobacteria were not affected, while the probiotic bacteria *Bifidobacterium breve* and *B. infantis* increased, along with short-chain fatty acids [54]. A different group also saw positive results in rats fed with a pomegranate extract [55]. So, it should be noted that the human body is an ecosystem that is comprised of human and bacterial cells, many of which help us digest our food [8]. When working properly, the gastrointestinal (GI) tract digests food, absorbs nutrients and expels waste. It is an important part of our immune system, as it distinguishes between pathogenic and helpful bacteria. It is also part of the endocrine system, as it secretes hormones that affect appetite, metabolism and body mass index (BMI). This includes the polypeptide hormones leptin, adiponectin, resistin and visfatin.

So, there have also been some important reports published after the 2012 review [42] that describe anti-obesity effects of pomegranates [56-59]. One of them described how oleanolic acid (1 – 25 μM) can block the production of visfatin in cultured adipocytes by disturbing a crucial signaling pathway [56]. That is, for visfatin (also known as pre-B colony-enhancing factor and nicotinamide phosphoribosyl transferase) to have its pro-inflammatory effects, it must activate the interleukin-6 (IL-6), tumor necrosis factor receptor associated factor 6 (TRAF 6) and nuclear factor-κB (NF-κB). So, oleanolic acid and foods like pomegranates that contain it may be effective against adipogenesis and inflammation caused by visfatin [56].

Visfatin can bind to the insulin receptor at a different (allosteric) site than the one to which insulin binds [60]. It causes the liver to release less glucose, while stimulating adipocytes and myocytes (muscle cells) to use more glucose. This leads to hypoglycemia. In some models of obesity, visfatin is upregulated [60]. It works together with leptin and resistin to cause insulin resistance, type-2 diabetes and atherosclerosis [61]. The concentration of leptin in blood serum depends on the BMI [61]. Leptin and insulin work together to regulate food intake and energy production by binding to their respective receptors. Leptin has many other effects, including stimulating the growth of new blood vessels, or angiogenesis, in metastatic breast cancer [62]. Resistin was named for its resistance to the effects of insulin [61]. It is also a pro-inflammatory hormone that plays an important role in obesity and diabetes [61]. Visfatin also plays important roles in cardiovascular and periodontal diseases

[63, 64]. The possible benefits of pomegranates to heart and oral health will be discussed shortly.

For now, it should be noted that the production of resistin in cultured adipocytes was also inhibited by oleanolic acid [57]. It did this by suppressing signaling though signal transducer and activator of transcription factor (STAT) 1, STAT3 and tyrosine kinase 2, or Tyk2 [57]. It also activated suppressor of cytokine signaling 3 (SOCS3), which inhibits Tyk2 activity [57]. STAT 1 and 3 are transcription factors that activate the transcription of genes coding for pro-inflammatory cytokines, such as interleukins and TNF-α (tumor necrosis factor- α) [8]. Tyk 2 is also known as Janus tyrosine kinase, or JAK [8]. It is a non-receptor tyrosine kinase that catalyzes the attachment of a phosphate to downstream proteins involved in cellular communication and diseases.

Fat cells can also be problematic in the liver. That is, non-alcoholic fatty liver disease can be caused by mitochondrial dysfunction in obesity [58]. "A commercial extract of pomegranates (150 mg/kg per day) that contained 40% punicalagin inhibited hyperlipidemia in rats on a high fat diet. The adipokines leptin and adiponectin were elevated in controls along with serum triglycerides and cholesterol, but not in rats that were given the pomegranate extract. The extract also increased the healthy high density lipoproteins (HDLs). It promoted mitochondrial function while eliminating oxidative stress and inflammation" [58].

Fatty liver can also be caused by drinking too much alcohol. Some of the harmful effects may be reduced by pomegranates, though. That is, a methanolic extract of pomegranate peels (200 mg/kg BW for 30 days) was shown to reduce the harmful effects caused by ethanol-induced fatty liver in rats [59]. That is, ethanol increased the concentrations of enzymes that indicate liver damage (bilirubin, alanine transaminase or ALT, aspartate aminotransferase or AST, and alkaline phosphatase or ALP) as well as total cholesterol, triglycerides, HDL-cholesterol, LDL-cholesterol and VLDL-cholesterol. This was prevented by the peel extract. It was prepared by soaking about 100 g of dried peels in 1 L of methanol for 5-7 days, with manual stirring every 24 hrs [59]. So, even better results could be expected if pressurized liquid extraction were to be used.

## Gastrointestinal Tract

Obesity and smoldering inflammation can also damage the gastro-intestinal (GI) tract. The anti-inflammatory effects of pomegranates on the GI tract were reviewed recently [65]. For example, smoldering inflammation can make the GI tract more susceptible to infection by the bacterium *Helicobacter pylori*, which can lead to ulcers and stomach cancer [8]. A methanol extract of pomegranate peels was shown to be as effective as the popular antibiotic metronidazole in inhibiting the growth of *H. pylori* [65, 66]. Other studies found that a methanolic extract of the peel also inhibited *H. pylori* growth [66, 67, 68]. One of them [66] did not tell how the methanolic extraction was done, but the other did – it used 80% methanol in a Soxhlet extraction for 4 hrs [67]. In the other study, dried pomegranates peels from Thailand were extracted with either water or 95% ethanol at room temperature for seven days [68]. Both extracts inhibited the growth of *H. pylori*, but the ethanolic extract worked better [68]. However, there is no way to know if this method or the Soxhlet extraction method can solubilize more bioactive materials. In a Soxhlet extraction, the solvent is continuously boiled and re-condensed, so the extraction temperature would have been that of boiling 80% methanol. Presumably the other 20% was water. If so, sugars and most phenolic compounds would have been in the extract. However, heat and methanol denature or destroy proteins. Also, lipophilic compounds, such as triglycerides, terpenoids and plant sterols would not have been solubilized. So, it is quite likely that either the entire peel (as a food additive) or an extract prepared by using dry methanol or dry ethanol at high temperature and pressure would be even more effective.

Another study showed that an extract of peels could inhibit the formation of ulcers caused by ethanol consumption in rats [65, 69]. They extracted dried peels (rinds) from India with 70% methanol at room temperature for 24 hrs [69]. This will only solubilize a portion of the hydrophilic active ingredients and none of the hydrophilic compounds. Still, the extract was able to decrease mucosal injury in the GI tract that was caused by either ethanol or aspirin [69]. It did this by preventing damage to important natural antioxidants, the enzymes superoxide dismutase (SOD) and catalase as well as the tripeptide glutathione [69]. It also reduced lipid peroxidation [69]. Still, it is quite likely that either whole peels or a more efficient extraction method would work even better.

In another study, an 80% methanolic extract of dried Indian pomegranate flowers to protect rats from experimentally-induced ulcers [70]. Although the title of the paper said that a "standardized extract" was used, although there is no widely recognized standard, in contrast to standardized extracts of other foods and herbs, such as *Ginkgo biloba* [71], which is has definite quantities of unique compounds, including ginkgolides A, B, C and trilactonic sesquiterpene bilabolids [70]. They did say that they solubilized 81.61 g of material by doing a Soxhlet extraction on 200 g of dried flowers for 12 hrs using 80% methanol, resulting in a 40.8% yield [70]. They also said that they obtained a 36.265% yield with hot alcohol [70]. Regardless of how the extraction was done, at least one of the extracts decreased the severity of lesions by 66% and the volume of gastric acid by 40% at a dose of 980 mg/kg body weight [70]. The extract also had a prophylactic effect in preventing the formation of excess gastric acid [70]. Still, whole flowers or a more efficiently prepared extract would probably have the same or even better effects.

Pomegranate tannins were able to inhibit ethanol-induced mucosal injury due to their ability to modulate the activities of superoxide dismutase (SOD) and glutathione peroxidase [65]. In another study, an aqueous extract of peels prevented ethanol-induced gastric ulceration in rats [72]. That is, 3 g of dried peels were extracted with 100 mL of boiling distilled water for 10 min [72]. This extract had 3.67 mg/mL total phenolic compounds and 4.41 mg/mL tannic acid [72].

Ellagic acid, an important compound in pomegranates and some other fruits, at doses of 3, 10 and 30 mg/kg BW exhibited anti-ulcer effects in animal models of acute and chronic gastric ulceration caused by ethanol and acetic acid, respectively [73]. It did this by increasing the endogenous production of nitric oxide (NO), replenishing non-protein sulfhydryls and limiting the production of pro-inflammatory TNF-α, interferon-γ and two interleukins (IL-4 and IL-6) [65]. That is, NO can protect against mucosal damage by acting as a vasodilator and potent inhibitor of leukocytes adherence to the vascular endothelium [74]. Moreover, non-protein sulfhydryls, like the one in glutathione, are potent natural antioxidants that help control the release of pro-inflammatory cytokines, while detoxifying reactive oxygen substances [65]. So, maintaining them can help prevent ulceration [65].

Aspirin (acetylsalicylic acid) and other non-steroid anti-inflammatory drugs (NSAIDS) can cause gastritis and internal bleeding by inhibiting the enzymes cyclooxygenase-1 (COX-1) and COX-2 [8]. A 70% methanol extract of dried rinds reduced the ulcer index of rats that had been given 400 mg/kg BW acetylsalicylic acid [69].

Another animal model is to ligate the pylorus [75]. Pomegranate tannins increased the secretion of adherent mucus and free mucus, but did not affect the free acidity, total acidity, gastric juice volume, and gastrin pepsin activity induced by pylorus ligation [65, 75]. One study used 50% ethanol to extract the peels and seeds [76]. The extracts reduced mucosal injury in this animal model. Another study on hogs [77] showed that ellagic acid inhibits the gastric $Na^+$,$K^+$-ATPase, enzyme that catalyzes the production of acid ($H^+$) [8]. A concentration of 2.1 μM caused a 50% inhibition [77]. Moreover, intraperioneal administration of 5 mg/kg DW ellagic acid reduced the incidence of gastric lesions [77].

Another animal model uses ischemia (lack of blood flow), followed by perfusion to cause gastric damage and nearly 50% reduction in gastric mucosal blood flow [78]. Ellagic acid (0.1 – 10 mg/mL) in the reperfusion fluid prevented the formation of lesions by suppressing lipid peroxidation [78].

Pomegranates and/or extracts prepared from them can also prevent intestinal inflammation [65]. One group showed that pomegranate juice, (6-50 μg/mL), total pomegranate tannins (30-200 μg/mL) and punicalagin (25-200 μg/mL) can suppress pro-inflammatory cell signaling in colon cancer cells grown in cell culture [79]. The juice suppressed COX-2 activation induced by TNF-α, reduced phosphorylation of the p65 subunit and binding to the NFκB and eliminated activation of protein kinase B [79], also known as phosphatidylinositol 3-kinase and PI3K [8]. That is, PI3K activates Akt, which catalyzes the phosphorylation and activation of the IκB kinase, which activates NFκB [8, 79]. PI3K and Akt are important in insulin signaling, breast cancer and the immune response [8]. There are also prescription anticancer drugs (sirolimus, temsirolimus and ridaforolimus) that target the PI3K/Akt/mTOR signaling pathway, where mTOR is the mammalian target of rapamycin [8]. There are many clinical tials being conducted on new chemical entities that target the PI3K-Akt pathway [8, 80]. Finally, it was shown that more than one compound or class of compounds in pomegranate juice is responsible for these health effects – possibly acting synergistically [79].

In a different cellular model (human intestinal Caco-2 cells) a hot (80 °C) aqueous extract of pomegranate peels reduced the concentrations of many mediators of cellular inflammation, including IL-8 and IL-1β [81]. It also inhibited the activation of NFκB and phosphorylation of the extracellular regulated kinase, which is part of a pathway (Ras-Raf-MEK-ERK) that contains several oncogenes, is deregulated in about 30% of all human cancers and has emerged as a target for anticancer therapy [82]. The Caco-2 cells come from cancer cells in the large intestine. When grown in a confluent monolayer, they differentiate into a polarized epithelial cell monolayer. This is used throughout the pharmaceutical industry to model the small intestinal mucosa so that the absorption of orally administered drugs can be evaluated [83]. The hot aqueous extract contained 108 mg GAE/g DW and an ORAC antioxidant capacity of 1068 μmol trolox equivalents per g DW [81]. These results indicate that this extract may be able to inhibit intestinal inflammation. Moreover, whole peels or more efficiently extracted peels could have a much bigger effect.

Subsequent studies on a "hydroalcoholic extract" of pomegranate peels (husks) and purified punicalagin showed that they can increase the transcription of genes coding for the pro-inflammatory ineterleukins IL-6 and IL-8, as well as monocyte chemoattractant protein, MCP-1 in Caco-2 cells [84]. They also increased the amounts that were secreted, thus displaying potent anti-inflammatory activities [84].

Endogenous bacteria in the gut can also contribute to intestinal inflammation and autoimmune diseases [8]. One study showed that a commercially prepared extract (PomX) and punicalagin selectively killed pathogenic bacteria and not probiotic bacteria grown in cell culture [54]. In general, phenolic compounds (especially hydrolyzable tannins) influence gut bacteria [85], and inhibit the growth of pathogenic bacteria [86]. Tannins can remain in the gut for at least 48 hrs [87]. While in the gut, they can inhibit bacterial growth by forming stable, non-covalent complexes with proteins, starch and essential metals [88]. They are present at a concentration of about 2 g/L in juice and are enriched in an extract called POMx [54]. Its composition was 19% ellagitannins (as punicalagins and punicalins), 4% free ellagic acid, and 77% oligomers (2-10 repeating units of gallic acid, ellagic acid, and glucose in different combinations) [54]. When added to growth media at a concentration of 0.01%, POMx inhibited the growth of potentially pathogenic *Staphylococcus aureus*, *Clostridium perfringens* and *C. clostridioforme* [54]. It was estimated that eating one pill of POMx would be the equivalent of having a concentration of 0.01% POMx in the gut, while drinking 500 mL of juice would provide a concentration of about 0.05% punicalagins to the colon [54]. Moreover, individual components of pomegranate juice and POMx (punicalagins, ellagic acid and gallic acid) inhibited the growth of *S. aureus*, *C. perfringens* and *C. clostridioforme*, too [54]. They all did this without inhibiting the growth of probiotic bifidobacteria and lactobacilli species [54].

This line of research was continued in a subsequent study that found that POMx increased the growth of probiotic bacteria by being metabolized into a class of compounds called urolithins [89]. For example, urolithin A is shown in Figure 1.

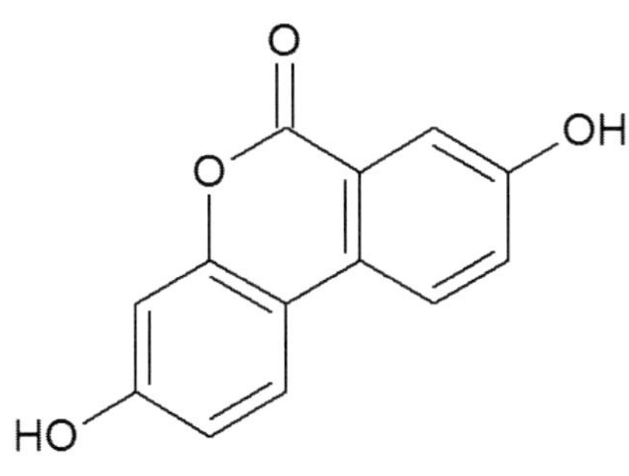


Figure 1. Structure of Urolithin A.

They found that the enhanced growth of probiotic bacteria was probably due to polyphenol oligomers made up of gallic acid, ellagic acid and glucose units (or –mers) [89]. They also found that POMx increased the production of acetate, propionate and butyrate, which are often called short chain fatty acids [89]. They can help to prevent cancer and help convert cholesterol into bile acids [90]. It is also possible that the anti-obesity properties of pomegranates and products made from them may be due, at least in part, to the stimulation of probiotic gut bacteria [89].

Ellagic acid (50 μM) downregulated the transcription of some of the genes in inflamed Caco-2 cells [91]. These genes code for proteins involved in intestinal inflammation, including signal transducer and activator of transcription-3, or STAT-3. It is a transcription factor that activates the transcription of NF-κB [91]. In another study, ellagic acid, urolithins A and B and a mixture of them, at concentrations achievable in the diet (10 and 40 μM for ellagic acid and urolithins, respectively), inhibited the growth of Caco-2 cancer cells by deactivating the extracellular signal-regulated protein kinase (ERK) signaling pathway [92]. The mixture altered the expression of genes that code for proteins that help control the cell cycle [92]. This includes several cyclins and cyclin dependent kinases [92]. ERK is also known as mitogen activated protein kinase, or MAPK [8]. Inhibition of MAPK/ERK may be able to not only prevent inflammation, but also colon cancer [91]. In subsequent study, urolithins were shown to prevent inflammation in Caco-2 cells that were stimulated with pro-inflammatory NF-κB [93]. Urolithin-A downregulated the expression of genes coding for the pro-inflammatory enzymes COX-2 and microsomal prostaglandin $E_2$ ($PGE_2$) synthase-1 [93]. This also prevented the activation of NF-κB and MAPK [93].

A mixture of ellagic acid together with urolithins A and B was able to improve the inflammatory responses of colon fibroblasts that were stimulated by IL-1β [94]. The mixture prevented the production of prostaglandin $E_2$ ($PGE_2$) that was induced by IL-1β in these cells and plays a key role in intestinal inflammation. The mixture also downregulated the expression of the gene coding for plasminogen activator inhibitor-1 (PAI-1), which is also important in inflammatory reactions in the intestine mucosa and is required for fibroblast migration. This prevented the effects of platelet-derived growth factor (PDGF), which is also needed for cell migration [94]. Finally, the mixture also inhibited the secretion of IL-8 and decreased the concentrations of intercellular adhesion molecule-1 (ICAM-1) and vascular cell adhesion molecule-1 (VCAM-1) [94], which mediate the adhesion of various white blood cells to the epithelium.

Ulcerative colitis is another form of intestinal inflammation. Ellagic acid at doses of 1-10 mg/kg BW reduced the severity of colonic lesions [95]. This was probably due to its antioxidant effect, since the activity of enzyme myeloperoxidase (MPO) was lowered and there was less lipid peroxidation in the colonic mucosa of test animals [95]. In a subsequent study, a hydroalcoholic extract (75% methanol with 25% water) of pomegranate flowers that contain ellagic acid also reduced the activity of MPO and the levels of oxidative markers in mice who had ulcerative colitis [96]. Moreover, this extract stabilized mast cells, which release many proinflammatory molecules, including histamine [96].

To help understand the anti-inflammatory effects of ellagic acid (60 mg/kg BW for 30 weeks) in the gut, its effect on colon carcinogenesis was studied [97]. It reduced inflammation while decreasing the biosynthesis of COX-2, inducible nitric oxide synthase (iNOS), IL-6 and TNF-α by inhibiting the NF-κB system [97]. So, it may be quite useful in preventing inflammation that can lead to colon cancer. This mechanism was also seen when ellagic acid (10-20 mg/kg BW) prevented colitis in rats [95, 96]. It was also shown that ellagic acid can reduce the infiltration of neutrophils (a type of white blood cell) [98]. It also increased the production of mucus in globet cells. All these anti-inflammatory effects might be due to reduced expression of COX-2 and iNOS [98].

Another study analyzed the effect of ellagic acid in a model of chronic colitis [99]. Consumption of a commercial PG fruit extract (250–500mg/kg) with or without being enriched with ellagic acid (10mg/kg) reduced inflammatory symptoms and histological injuries in rats suffering from chronic colonic inflammation [99]. It also decreased MPO activity and TNF-$\alpha$ production, while reducing COX-2 and iNOS expression by inhibiting MAPKs and NF-$\kappa$B [99].

In a rat model of ulcerative colitis, an aqueous extract of pomegranate peels (200–800mg/kg) relieved diarrhea and ulcers, while decreasing MPO activity, lipid peroxidation, IL-1$\beta$, and TNF-$\alpha$ concentrations [100]. Another study showed that a peel extract could downregulate the expression of genes coding for the pro inflammatory proteins COX-2, IL-6, and IL-1$\beta$ in colonic and adipose tissues from obese mice who consumed a high-fat diet [101]. The same extract also exhibited a probiotic effect by increasing bifidobacteria [101], which are bacteria showing high anti-flogistic activity (reduces fever and inflammation) in the intestinal tract [102].

In a subsequent study, a rat model of colitis was used to assess whether antiflogistic properties seen with pomegranate peel extract were due to ellagitanins or to their urolithin metabolites [103]. Rats were fed with the peel extract (250mg/kg) or urolithin A (15mg/kg) for 25 days before inducing colitis. The rats fed the extract and urolithins, the levels of bifidobacteria and lactobacilli increased [103]. They also increased the count of *Clostridium* probiotic strains and maintained it after inducing colitis. This prevented the colonization and invasion of colonic tissue by pathogenic enterobacteria [103]. Moreover, the metabolic fate of the extract after being metabolized by gut bacteria was different in rats with colitis compared to those without colon inflammation [103]. That is, the feces of rats who had inflamed colons contained ellagic acid and punicalagin, but only urolithins were found in the feces of healthy rats [99]. So, the lower metabolic activity of gut microbiota in an inflamed colon is important when studying the effects of phenolic compounds [65]. The anti-inflammatory activity of pomegranates and products made from them in the model for inflammatory bowel disease may be explained by the synergistic activities of urolithins [103].

Pomegranate seed oil has also been shown to be able to decrease ulcerative damage in a rat model of colitis [104]. As mentioned in chapter 4, the seed oil is almost 100% triglycerides that contain fatty acyls attached to the glyceride backbone. The most abundant fatty acyl is punicoyl. So, when triglycerides from pomegranate seed oil are digested, punicic acid is formed, along with other fatty acids and glycerol. The effectiveness of oil and commercially available pure punicic acid was evaluated in rats whose colitis was induced by hyperactivated neutrophils that were induced by tumor necrosis factor-α, or TNF-α [104]. The neutrophils produce reactive oxygen substances after TNF-α causes the enzyme NADPH oxidase to be phosphorylated on one of its subunits (p47phox) in a reaction catalyzed by p38 MAP kinase. This is important in inflammatory bowel diseases including ulcerative colitis, Crohn's disease and infectious enterocolitis. Punicic acid inhibited this phosphorylation, as did the seed oil. It also inhibited the release of MPO, TNF-α of reactive oxygen substances and the release of degradative enzymes that are important in inflammatory bowel diseases [104].

Punicic acid may also be able to prevent inflammatory bowel diseases [105]. It upregulated the activation of peroxisome proliferator-activated receptor-γ (PPAR-γ) and suppressed the expression of genes coding for TNF-α and MCP-1. This was in agreement with previous studies that showed that PPARs are important targets of dietary lipids [106]. Pomegranate seed oil was also able to reduce the incidence and severity of necrotizing enterocolitis by protecting the epithelial barrier and preserving the intestinal integrity [107]. "This disease is the major cause of morbidity and mortality in premature infants" [107].

Diarrhea is a major symptom of inflammatory bowel diseases [65]. So, it is noteworthy that an intraperitoneal injection of an aqueous extract of pomegranate peels [108] and a methanolic (7:3 $CH_3OH:H_2O$, v/v) extract [109] of peelss reduced diarrhea in a mouse model. The aqueous extract reduced intestinal weight and motility, while inhibiting the spontaneous ileum contractions induced by acetylcholine [108]. The methanolic extract had a strong antioxidant effect [109]. It scavenged NO, which is a reactive oxygen substance and second messenger [8] that is involved in causing diarrhea [109]. The methanolic extract was prepared by Soxhlet extraction of an unspecified amount of dried peels with 70% methanol plus 30% water for an unspecified time [108]. The aqueous extract was prepared by stirring a mixture of 150 g of dried peels with 3 L of boiling water [109]. Neither method is very efficient. So, whole peels (or rinds) may be much more effective – especially if used as a food additive. Moreover, extraction with dry methanol or ethanol at elevated temperature (100 °C) and pressure (100 atmospheres) will extract much more material, while preventing the hydrolysis of bioactive anthocyanins.

## DIABETES

Since pomegranates and products made from them can help prevent smoldering inflammation and obesity, they can also help prevent and treat diabetes. As mentioned previously, they contain many phenolic compounds that can chelate $Fe^{2+}$ and $Cu^{2+}$ and prevent the formation of highly toxic hydroxyl radicals formed by the Fenton reaction [15, 16]. Even though this may be the most important mechanism by which pomegranates prevent diabetes and other diseases, there are others. Some of them were described in recent reviews [110, 111]. However, pomegranate seeds and flowers were used in India and other parts of the world to treat diabetes long before the advent of modern science and medicine [106]. Modern science has supported this, based on studies *in vitro* and *in vivo*, in animals and people [110, 111]. For example, a pomegranate flower extract was able to decrease heart damage in diabetic rats [112]. The extract was prepared by mixing 5 L of methanol with dried, powdered flowers five times at room temperature [112], so only a small portion of the bioactive ingredients were solubilized and given to the diabetic rats. Still, the extract (500 mg/kg) was able to limit cardiac fibrosis by activating PPAR-α and PPAR-γ [108]. This makes sense because the prescription drug pioglitazone (Actos®), is an agonist of PPAR-γ, which is important in maintaining blood glucose levels [8]. The extract also improved hyperglycemia and hyperlipidemia while suppressing the expression of cardiac fibronectin and collagen I and III mRNAs that were overexpressed in diabetic rats that did not get the extract [112]. Diabetic rats also had upregulated expression of mRNAs coding for endothelin-1 (ET-1), the type A receptor of ET-1 ($ET_A$), inhibitor κBβ and c-*jun* as well as downregulated inhibitor κBα. The pomegranate flower extract reduced these effects. It also inhibited the activation of macrophages, as did three of its bioactive ingredients, oleanolic acid, ursolic acid, and gallic acid [112].

ET-1 is a peptide that causes vasoconstriction and blood platelet aggregation [8]. In heart disease, vascular homeostasis is disrupted, which leads to further inflammation and a pro-thrombotic state. Reactive oxygen and nitrogen substances (RONS) like the superoxide ($O_2^-$) and peroxynitrate ($ONOO^-$) anions activate protein kinase C (PKC), which cause more RONS to be produced, in a viscious circle (positive feedback). RONS also cause vascular inflammation and upregulate NF-κB and activate the transcription of pro-inflammatory genes encoding for monocyte chemoattractant protein-1 (MCP-1), selectins, vascular cell adhesion molecule-1 (VCAM-1), and intracellular cell adhesion molecule-1 (ICAM-1). This helps monocytes stick to the vascular endothelium, leading to the formation of foam cells. Adhesion molecules stay up-regulated by the secretion of IL-1 and TNF-α from active macrophages. PKC also stimulates the biosynthesis of ET-1 [8].

NFκB can induce the biosynthesis of fibronectin in endothelial cells when glucose levels are too high [112]. Its activity is controlled by inhibitors κBβ and κBα. [8]. Diabetes increases the expression of mRNA that codes for fibronectin, but this requires NFκB [108]. There is also an activator protein-1 (AP-1) that is a heterodimer of c-jun and c-fos activated in diabetic rats. It is involved in fibronectin synthesis [112]. This helps explain why it is important that pomegranate flower extract affected the expression of mRNAs coding for ET-1, $ET_A$ and c-*jun*, as well as inhibitors κBβ and κBα [112]. Still, no extract will contain all of the bioactive ingredients in flowers or almost any other part of pomegranates. So, it is noteworthy that one study used whole flowers, as they are used in *Unani* and *Ayurvedic* medicine [113]. They gave them to mice and found that they reduced blood glucose concentrations and insulin resistance caused by aging [113], which is a big risk factor for type-2 diabetes. Moreover, whole flowers improved skin abnormalities such as excessive peri-ovary fat mass, decreases in skin water content, epidermis thickness, and collagen density in corium [113].

In another study, a methanolic extract of pomegranate flowers at a dose of 500 mg/kg per day increased the activity of PPAR-γ in a human cell line [114]. Others used 1:1 ethanol:water (v/v) to extract flowers and found that it lowered the blood glucose concentration in diabetic rats [115]. A subsequent study found that an extract of flowers prepared using methanol at room temperature was able to improve cardiac lipid metabolism in diabetic rats when given once daily for six weeks at a dose of 500 mg/kg [116]. That is, it reduced the concentration of triglycerides in the heart and blood plasma as well as fatty acids in the blood. It also prevented the overexpression of mRNAs coding for fatty acid transport protein, PPAR-α, carnitine palmitoyltransferase-1, acyl-CoA oxidase and 5'-AMP-activated protein kinase α2 [116]. It also restored the downregulated expression of mRNA coding for cardiac acetyl-CoA carboxylase [116]. The same group also found that flower extract increased the expression PPAR-γ mRNA and cardiac glucose transporter 4 in diabetic rats [110]. In a different approach, flowers were extracted with diethyl ether and the residue obtained healed wounds in diabetic rats [117]. A subsequent study showed that pomegranate flowers (500 mg/kg per day) decreased the amount of lipid peroxidation and increased the amount of reduced glutathione in diabetic rats, as well as improving their learning and memory performance [118].

There are also studies that looked at the antidiabetic effects of oleanolic acid [119-122]. It has antioxidant, microbicide, antidiabetic, anti-inflammatory, hypolipidemic, and antiatherosclerotic properties [119]. "It increases the biosynthesis and excretion of insulin as well as glucose tolerance. One way that it does this is to increase the release of the neurotransmitter acetylcholine from nerve terminals. It is also an agonist of the TGR5 receptor, which is also known as the G protein-coupled bile acid receptor 1 (GPBAR1) and membrane-type receptor for bile acids (M-BAR). That is, bile acids cause glucagon-like peptide-1 by acting on the TGR5 receptor. This increases the secretion of insulin and the regeneration of pancreatic β-cells. Oleanolic acid also protects β-cells from oxidative damage caused by $H_2O_2$ and other reactive oxygen species. It also increases the activity of the enzyme Src-homolgy phosphotyrosyl phosphatase 2 (Shp-2) that is important in insulin signaling and biosynthesis. It increases the response to insulin by restore the autophosphorylation of the insulin receptor, which activates it. It inhibits the protein tyrosine phosphatase enzymes PTP1B and TCPTP that suppress insulin signaling. It activates the phosphatidyl inositol 3-kinase (PI3K) pathway that is required for uptaking glucose and synthesizing glycogen. Another kinase, AMP kinase, is also activated by oleanolic acid. It participates in glucose uptake and the oxidation of fatty acids. Oleanolic acid also inhibits glycogen synthase kinase 3-β, which inhibits the effects of insulin. Moreover, it alleviates oxidative stress caused by insulin resistance. It also affects the immune system. That is, many inflammatory cytokines that decrease insulin signaling are regulated by the transcription factor NFκB. It is activated by stumili that cause an inhibitory subunit, IκB, to be phosphorylated in a reaction catalyzed by IKK. Oleanolic acid and some other terpenes inhibit IκB kinase (IKK), thus preventing NFκB from being activated. Oleanoic acid and ursolic acid inhibit another important enzyme, aldol reductase. This enzyme catalyzes the rate-limiting step in a metabolic pathway called the polyol pathway. Chronic hyperglycemia causes excess glucose to be shunted from the normally healthy glycolytic pathway to the polyol pathway that produces relatively high concentrations of sorbitol and fructose. These can form advanced glycation end products (AGEs). Oleanolic and ursolic acid help prevent this from happening. In addition, oleanolic acid has hypoglycemic affects and modulates PPAR activity. However, the main contribution of oleanolic acid to preventing diabetes is by interacting signaling transduction pathways that increase the transcription of key defensive genes, such as Nrf2, also known as nuclear factor-like 2. The Nrf protein for which this gene codes binds to antioxidant response elements and induces the transcription of cytoprotective genes that code for proteins that respond to oxidative stress. Nrf is also involved in the expression of enzymes that use NADPH, which is a natural reducing agent that increases the production of the endogenous anti-inflamamtory molecule, reduced glutathione" [119]. It should be noted that protein kinases catalyze the phosphorylation of other proteins [8].

Oleanolic acid has also been shown to attenuate insulin resistance in rat adipose tissue induced by fructose [120]. "Insulin resistance in adipose tissue causes abnormal metabolism by causing more free fatty acids to be released as triglycerides are hydrolyzed. Oleanolic acid was given to rats by oral gavage for 10 weeks at a dose of 25 mg/kg once daily. Insulin resistance was induced by giving one group free access to an aqueous solution of 10% fructose. Oleanolic acid attenuated the increase in the concentrations of insulin and free fatty acids in the blood. It also limited adipose tissue insulin resistance caused by excess fructose consumption. It also increased the expression of the insulin receptor substrate (IRS-1) and the enzyme phosphatidyl inositol 3-kinase (PI3K). It also increased the phosphorylation of the enzyme called Akt or protein kinase B" [120]. This enzyme is part of the PI3K/Akt/mTOR signaling pathway that is important in not just glucose metabolism, but also in cancer [8].

In another study, oleanolic acid was shown to improve glucose tolerance in normal mice and decrease visceral obesity in mice that were fed a high-fat diet [121]. Excess visceral fat is a major risk factor for diabetes and cardiovascular diseases [8]. Adult mice were given oral doses of oleanolic acid (5, 10 and 20 mg/kg p.o.) once daily for seven days [121]. "All doses had an effect, but the highest dose had the greatest effect. It decreased the weight of visceral fat. Oleanolic acid also decreased the activity of the enzyme amylase, which catalyzes the hydrolysis of starch. Thus, the digestion of starch was lower and can help reduce body weight. Oleanolic acid also decreased the concentrations of glucose, triglycerides and cholesterol in the blood. It also decreased the microvascularization of the liver, which is a measure of steatosis, or the abnormal retention of lipids" [121].

In another study, a methanolic extract of pomegranate seeds prevented hyperglycemia in diabetic rats at doses of 150 to 600 mg/kg BW [122]. It was even more effective than the prescription drug chlorpropamide [122]. Still, it should be noted that methanol will only extract relatively hydrophilic bioactive compounds like phenolics, but not lipophilic triglycerides and terpenes. Another study found that pomegranate seed oil could improve insulin sensitivity and adiponectin in diabetic mice [123].

Pomegranate peels may also be useful in helping to treat or prevent diabetes. One study showed that an aqueous extract of powdered peels can decrease blood glucose concentration while increasing insulin levels and regenerating pancreatic β-cells that biosynthesize insulin [124]. The extract was prepared by adding 200 g of boiling water to three grams of dried peels and leaving it for 10 min [124]. It was filtered and dried before giving it to diabetic rats for four weeks at a dose of 0.43 g of extract per rat [124]. This extraction method will only solubilize hydrophilic compounds, such as sugars and some phenolic compounds. It did not include dietary fiber, an important ingredient in peels that can help prevent obesity and diabetes [8]. So, the entire peel may be even more effective as a food additive. In another study, an aqueous extract of pomegranate peels inhibited the lipid peroxidation induced by hydrogen peroxide in rat red blood cells grown in cell culture [125].

Concentrated pomegranate juice containing 2.6 mg/mL GAE of polyphenols was beneficial to people who were diabetic [126]. "They were given 50 mL per day for four weeks. It decreased the concentrations of lipid peroxides and other reactive oxygen substances. It did this by stabilizing a protein called paraoxonase (PON1) that is associated with low density lipoproteins (LDLs). PON1 is an esterase, so it can catalyze the hydrolysis of esters. It breaks down oxidized lipids in oxidized lipoproteins and macrophages, as well as inhibits the uptake of oxidized LDLs by cells and decreases the biosynthesis of cholesterol in macrophages. It stimulates the efflux of cholesterol that is mediated by healthy HDLs. PON1 protects against atherosclerosis, but is partly depleted in diabetics. So, stabilizing PON1 is an important antidiabetic property of pomegranate juice. It is noteworthy that the juice was antidiabetic despite having about 10% total sugars, mostly as glycosides bound to phenolics. Moreover, a polyphenol extract that did not have a high sugar content was not as effective as the juice" [126]. This is another study that supports the idea that the whole juice is healthier than an extract. So, when glucose is covalently bound to phenolic compounds, it may have very different nutritional, relatively non-toxic properties than free glucose.

Another study [127] found that 40 g of pomegranate juice concentrate reduced total cholesterol, LDL cholesterol, LDL/HDL ratio, and total cholesterol/HDL ratio in patients with type 2 diabetes [127]. Another study found that pomegranate juice and one of its ingredients, punicalacgin made β cells release more insulin [128]. A more recent study showed that an extract of pomegranate juice and ellagic acid can reduce the secretion and intracellular concentration of the protein resistin (an adipokine) in differentiated mouse adipocytes [129]. Resistin is often called "the link between obesity and type-2 diabetes" [130]. The prescription drug rosiglitazone downregulates the expression of the *Retn* gene that codes for resistin [131]. People who have metabolic syndrome (obesity) and diabetes have higher concentrations of resistin in their blood [132]. There is an animal model (ovariectomized mice) in which resistin is elevated [130]. An extract of pomegranate fruits and an active ingredient (ellagic acid) were given to such mice. Both of them suppressed the secretion of insulin in them and in the mice and cultured adipocytes. It stimulated the degradation of resistin [130]. To prepare the extract, a commercial source of Japanese pomegranate juice was dialyzed through a cellulose tube into water overnight, followed by freeze drying to obtain a powder [130]. Its composition would be quite different than methanol or ethanol extractions. It would have all the ingredients of the juice except for proteins and any insoluble solids that would not pass through the cellulose membrane. Solvent extraction would not have solubilized the proteins either, but it would not have solubilized hydrophobic compounds such as terpenes that might be in the juice, especially if seeds were not removed before squeezing juice from the fruit. Also, solvent extraction at room temperature and pressure might not even be able to extract all of the hydrophilic compounds, including glucose, which is not very soluble in alcohol. So, it is noteworthy that the presence of glucose and fructose in pomegranate juice or its dialysate did not interfere with its antidiabetic behavior.

One of the common effects of obesity and diabetes is chronic kidney disease, which is diagnosed by elevated concentrations of creatinine and blood urea nitrogen (BUN) [8]. In a recent study, pomegranate juice was shown to be able to protect rabbits from kidney toxicity (nephrotoxicity) that was chemically induced by amikacin [133]. Amikacin is an aminoglycoside antibiotic that was designed to be effective against bacteria that are resistant to another antibiotic, streptomycin [8, 134]. However, it can cause nephrotoxicity [133, 135]. Since kidney damage is also accompanied by oxidative stress, it was thought that the antioxidants in pomegranate juice might help prevent nephrotoxicity [133]. So, it is important that pomegranate juice was able to reduce creatinine and BUN [133]. However, it is rather amazing that this article [133] said that they used an extract of pomegranate juice. What they actually reported was that they started by obtaining about 250 mL of juice by cold pressing about 2 kg of pomegranate fruits obtained in Iraq. Next, they concentrated the juice by distillation. Then they added an equal volume of ethanol, shook the mixture 12 hrs at room temperature, then evaporated off the ethanol. Finally, they removed the water by freeze drying and desiccating the remaining aqueous solution. However, they said that they gave rabbits an oral dose of "100 mL several times daily" ... "for 14 days" [133]. So, the actual dose was not specified. Moreover, ethanol and water mix at all proportions, so adding ethanol to an aqueous solution of concentrated juice will not extract anything. So, adding ethanol and then evaporating it off accomplished nothing. It should also be noted that important anthocyanins are subject to destruction by hydrolysis, so the first step of preparing the so-called "extract" (distillation or boiling) probably destroyed most of the bioactive anthocyanins. It certainly denatured (destroyed) any bioactive proteins that may have been present. So, one should not conclude from this study that commercial pomegranate extracts (which are prepared by a genuine extraction) will have the same effect. Finally, it is quite possible that pomegranate juice can have the same effects or even better effects in preventing nephrotoxicity, without the need for boiling, distilling or adding ethanol.

Pomegranates can also increase the flow of urine, which is called diuresis. At least seven ellagitannins in pomegranate pericarps (peels) inhibit the enzyme carbonic anhydrase [134], which catalyzes the following reaction:

$$CO_2 + H_2O \rightleftharpoons HCO_3^- + H^+$$

Like all enzymes, it catalyzes the forward as well as the reverse reaction, depending on the needs of the cell [8]. This is a reversible reaction that happens slower without carbonic anhydrase [135]. "Its physiological function is to maintain pH balance in the blood and urine. That is, when $H^+$ is secreted in the kidney tubules, it combines with filtered $HCO_3^-$ to form $H_2CO_3$. Carbonic anhydrase then catalyzes the formation of $CO_2$, which diffuses across cell membranes. $H^+$ and $HCO_3^-$ react to form $H_2CO_3$ so more $HCO_3^-$ can is continuously reabsorbed from the pre-urine. However, in the presence of the prescription drug acetazolamide, carbonic anhydrase is inhibited and the concentration of $H_2CO_3$ increases. This slows down the reaction between $H^+$ and $HCO_3^-$ and decreases the body's ability to reabsorb $HCO_3^-$ in the blood. This leads to a decreased ability to exchange $Na^+$ for $H^+$, resulting in mild diuresis. Prescription carbonic anhydrase inhibitors like acetazolamide (Diamox®) are used to treat glaucoma, mountain sickness, gastric and duodenal ulcers, neurological disorders and osteoporosis" [135]. Some of them are also used as anti-epileptic agents and diuretics, which increase the rate of urine formation. For example, acetazolamide (Diamox®) is used to treat glaucoma, idiopathic intracranial hypertension, altitude sickness and epilepsy [135]. Even though pomegranates have not been tested for their effects on glaucoma, mountain sickness, gastric and duodenal ulcers, neurological disorders or osteoporosis, it may have some effects on one or all of them.

## CARDIOVASCULAR DISEASE

As mentioned above, obesity and diabetes can lead to cardiovascular disease. However, there can be many other causes, such as aging, improper diet, metabolic syndrome (obesity), stress, lack of physical activity, high blood pressure, consuming tobacco products and poorly liganded $Fe^{2+}$ and $Cu^{2+}$ [8]. In contrast, consuming foods and beverages that contain relatively high amounts of phenolic compounds can help prevent cardiovascular diseases [136]. As we get older, vascular endothelial function can deteriorate. "Endothelial function, elastic properties of large arteries, and the magnitude and timing of wave reflections are important determinants of cardiovascular performance" [136]. So, it is important to look at studies on how pomegranates can prevent and/or treat high blood pressure and cardiovascular diseases.

The positive effects of pomegranate juice on high blood pressure (hypertension) and cardiovascular health have been reviewed [137]. In one study, hypertensive patients drank 50 mL of pomegranate juice containing 1.5 mmol total phenolic compounds every day for two weeks [138]. They enjoyed a 36% decrease in the activity of the angiotensin converting enzyme (ACE) and a 5% decrease in systolic blood pressure [138]. ACE catalyzes the conversion of angiotensin I to angiotensin II [139]. "This is part of the renin-angiotensin system (RAS), a hormone system that regulates blood pressure and water balance. When the blood volume is low, the kidneys activate prorenin and secrete renin into the blood plasma, where it catalyzes the conversion of angiotensinogen (released from the liver) to angiotensin I. It is then converted to angiotensin II. ACE is vasoactive and causes blood vessels to constrict, raising the blood pressure. Moreover, angiotensin II stimulates the adrenal cortex to secret aldosterone and the RAS/MAP pathway in the vasculature. This causes the tubules of the kidneys to increase the reabsorption of $Na^+$ and water into the blood. This increases the volume of fluid in the body, and thus the blood pressure. If the RAS is too active, it can cause hypertension (high blood pressure). There are many anti-hypertensive drugs that target the RAS. They are also prescribed to treat heart failure, kidney failure and to the harmful effects of diabetes" [139]. Please note that RAS is quite distinct from the oncoprotein ras, or the gene that codes for, *RAS*, which is an oncogene that can cause cancer when mutated [8].

High blood pressure and smoldering inflammation can lead to atherosclerosis and cardiovascular diseases [8]. Areas of the heart that are susceptible to atherosclerosis often have altered blood flow and perturbed shear stress [140]. This includes vascular endothelial cells that are in direct contact with the blood [140]. They are found on the surface of the heart and in the interior surface of the chambers of the heart [141]. They help control blood pressure [141]. Shear stress of endothelial cells causes nitric oxide (NO) to be released in a reaction catalyzed by endothelial NO synthase (eNOS) [140, 142]. When shear stress is perturbed, reactive oxygen and nitrogen substances (RONS) are produced, along with free radicals [140, 143]. Also, genes such as ELK-1 and *p*-JUN that are responsive to oxidative stress are expressed at higher levels [140]. This can lead to endothelial dysfunction and increased atherogenesis [140, 144]. So, researchers studied the effects of pomegranate juice on the expression of ELK-1 and *p*-JUN as well as the activity of eNOS in cultured human coronary artery endothelial cells that were subjected to shear stress and to mice who had high concentrations of cholesterol in their blood (hypercholesterolemic mice) [143]. "The juice was prepared by crushing and squeezing the 'Wonderful' variety of pomegranates, followed by adding the enzyme pectinase, which improved the extraction and filtration by preventing the formation of pectin gels. The juice was filtered, pasteurized and concentrated. It was then diluted with water (6.25 mL of concentrate diluted to 1 L). One group of 3-month old mice drank the juice for four weeks before being fed a high cholesterol diet. They continued drinking the juice for another 24 weeks. Another group were fed the high cholesterol diet for six months and then given the juice for 24 weeks. The mice drank an average of 5 mL per day, which was equivalent to 0.875 μmol of total polyphenols. It reduced the expression of *ELK-1* and *JUN* genes in mice and cultured endoethelial cells, while increasing the expression of the gene coding for eNOS. It also reduced the progression of atherosclerosis. Also, it reduced the size of atherosclerotic lesions when compared to control mice that did not drink the juice. The control mice also had more macrophage foam cells. So, pomegranate juice may help prevent atherosclerosis and reduce it after it occurs [143].

Pomegranate juice contains tannins that are anti-atherosclerotic [145]. It may have "anti-atherosclerotic, anti-hypertensive, antioxidant and anti-inflammatory effects in human subjects and mouse models" [145]. The portion of the juice that contains sugars and glycosides was shown to reduce the oxidative state of macrophages [145]. This is important, because when macrophages in the heart become oxidized, they can become foam cells that lead to atherosclerosis and cardiovascular diseases [8]. The sugars in pomegranate juice were separated from other components by preparative scale liquid chromatography [145]. "An octadecyl silica ($C_{18}$) sorbent was used to isolate the sugars, anthocyanins and tannins. The $C_{18}$ sorbent binds hydrophobic compounds, but hydrophilic sugars and some glycosides (sugars covalently attached to phenolic compounds) do not bind to it. This portion of the juice was given to diabetic mice as well as macrophages and cells that are similar to macrophages (J-774A.1) that were grown in cell culture. The activities of paraoxonases 1 and 2 (PON1 and PON2) in blood serum were assayed as well as total peroxide and glutathione concentrations. The free radical scavenging activities of purified delphinidin-3-O-glycoside and cyanidin-3-O-glycoside were also measured. It was shown that pomegranate juice can decrease the oxidative state of macrophages more than the phenolic compounds in the juice. The sugar fraction of the juice was able to decrease the oxidative stress of macrophages. Glucose increased and fructose decreased the oxidative stress, but the effect of fructose was small. The sugar fraction decreased the oxidative stress of macrophages in not just control (non-diabetic) mice, but also diabetic mice. In J-774A.1 cells in culture, the concentration of peroxide decreased by 40% and increased glutathione by 20%. It also reduced the peroxide concentration by 50% and increased glutathione by 12% in peritoneal macrophages isolated from diabetic mice. Moreover, when diabetic mice drank the sugar fraction of pomegranate juice (containing 30 mg of glucose per mouse) for 10 days, the concentration of peroxide decreased by 10% while glutathione decreased by 7%. The sugar fraction also decreased PON2 activity in macrophages of diabetic mice *in vivo* by 7% and by 16% in J-774A.1 cells. On the other hand, the activity of PON1 was not affected. Finally, delphinidin-3-O-glycoside and cyanidin-3-O-glycoside had relatively high antioxidant capacities" [145]. In conclusion, these results are important because they show that pomegranate juice may be able to help prevent cardiovascular diseases as well as stroke in not just normal mice, but also diabetic mice, despite the fact that it contains glucose and fructose. It is also quite interesting that the concentration of fructose in pomegranate juice was not harmful [145], even though highly concentrated fructose in sweetened beverages is quite harmful [8]. This is just one of many examples that show that in toxicology, "the dose is the poison" [8].

In a similar study, pomegranate juice (100 mg/kg and 300 mg/kg; p.o. for 4 weeks) was found to prevent an increase in arterial high blood pressure and changes in vascular reactivity in rats in which diabetes was caused by administering angiotensin II [146]. "It also reduced the concentrations of thiobarbituric acid reactive substances in their kidneys and pancreas and increased the activities of the important antioxidant enzymes superoxide dismutase (SOD), catalase, and glutathione reductase. That is, the juice prevented the increase in blood pressure caused by angiotensin II in diabetic rats by fighting the oxidative stress and inhibiting the activity of the angiotensin converting enzyme, or ACE" [146].

Flavonoids in pomegranate juice have been shown to help prevent the formation of atherosclerotic lesions and cardiovascular diseases in atherosclerotic mice and in humans [147]. "They prevented the oxidation of low density lipoproteins (LDL) and the formation of foam cells from macrophages. This is important because foam cells can oxidize LDLs. Flavonoids can bind to LDLs, preventing them from being oxidized by reactions catalyzed by enzymes in foam cells. The juice also increased the activity of paraoxonase, which caused lipid peroxides to be hydrolyzed in LDLs and atherosclerotic lesions" [147].

Even though these studies [141-147] showed that drinking pomegranate juice for a short time can help prevent cardiovascular diseases for a while, many people consume it for many years. This was addressed by a study in which juice was consumed for three years by patients who had vascular surgery in Israel [148]. "Pomegranates were crushed and squeezed to obtain a juice that was then pasteurized, concentrated and stored at -18 $^{o}$C until used. The concentrate was diluted 1:5 with water to obtain a juice that had 1979 mg/l of tannins (1561 mg/L of punicalagins and 417 mg/l of hydrolyzable tannins), 384 mg/l of anthocyanins (delphinidin 3,5-diglucoside, cyanidin 3,5-diglucoside, delphinidin-3-glucoside, cyanidin 3-glucoside, pelargonidine 3-glucoside), and 121 mg/l of ellagic acids derivatives", as well as 3 mg of vitamin C per 100 ml of juice" [148]. In control patients with severe carotid artery stenosis who did not receive pomegranate juice (but did receive prescription statins, ACE inhibitors, calcium channel blockers and/or β-blockers) the mean intima media thickness

(IMT) increased by 9% in the first year of the study [148]. On the other hand, the IMT of patients who drank 50 mL of the juice "was reduced after 3, 6, 9 and 12 months of PJ consumption by 13%, 22%, 26% and 35%, respectively, in comparison to baseline values" [148]. The juice also reduced the mean peak systolic velocity "of both left and right carotid arteries by 21%" [148]. The mean carotid end diastolic velocity "of both left and right carotid arteries decreased, by 16%, 20%, 31% and 44%, after 3, 6, 9 and 12 months of PJ consumption, respectively" [148]. Also, "the patient's systolic blood pressure was significantly ($P<0.05$) reduced by 7%, 11%, 10%, 10% and 12% after 1, 3, 6, 9, and 12 months of PJ consumption, respectively, compared to values obtained before treatment" [148]. "The oxidative state of the blood serum of patients who received pomegranate juice (PJ) also decreased. This was indicated by a reduction in oxidized LDLs and an increase in the total antioxidant status and PON1 activity. There was also a decrease in LDL oxidation induced by $Cu^{2+}$. Moreover, the body mass index (a measure of obesity) did not change, nor did the concentrations of glucose and total lipids in blood serum. In one of the patients, the concentration of cholesterol in carotid lesions was reduced by 58% and another was reduced by 20%. The concentrations of lipid peroxides in the lesions were also reduced by 61% in one and 44% in the other. The concentration of the important endogenous antioxidant reduced glutathione was increased in the lesions by 43% and 32%. Finally, the juice was not toxic, as indicated by blood chemistry analyses of liver, heart and kidney functions" [148].

Not just the juice, but the arils from which it comes can help prevent cardiovascular disease, diabetes, osteoporosis, arteriosclerosis, and Alzheimer's disease, as part of the aging process [149, 150]. "Arterioscelosis occurs when the blood vessels that carry oxygen and nutrients from your heart to the rest of your body (arteries) become thick and stiff - sometimes restricting blood flow to your organs and tissues. Healthy arteries are flexible and elastic, but over time, the walls in the arteries can harden, a condition commonly called hardening of the arteries.

Atherosclerosis is a specific type of arteriosclerosis, but the terms are sometimes used interchangeably. Atherosclerosis refers to the buildup of fats, cholesterol and other substances in and on arterial walls (plaques), which can restrict blood flow. These plaques can burst, triggering a blood clot. Although atherosclerosis is often considered a heart problem, it can affect arteries anywhere in the body" [151].

Ellagitannin oligomers and a neolignan in arils inhibited the formation of advanced glycation end products (AGEs) *in vitro* [149]. "That is, the oligomers and neolignan were each added individually to a phopshate buffered aqueous solution containing 0.8 mg/mL of the protein human serum albumin and 0.2 M (moles/L) of glucose. Ellagic acid and urolithins A and C were also tested. Pomegralignan and the two urolithins had little or no inhibitory activity, but eight other compounds did. The concentrations needed to cause 50% inhibition ($IC_{50}$) ranged from 0.03 to 0.37 μM, with pomegraniin A being the most active (lowest $IC_{50}$) and oenothein B being the least active" [149]. This is important because elevated concentrations of glucose and its metabolites (such as methyl glyoxal) in the blood serum can react with proteins and tissues in the body [8]. AGEs and reactive oxygen and nitrogen species (RONS) are toxic. They bioaccumulate and lower the sensitivity of tissues to insulin. High levels of glucose can also inhibit the immune system, making one more susceptible to many diseases, including cancer [8]. "In fact, beverages sweetened with high fructose corn syrup have been called public health enemy number one in the USA, since they not only have too much sugar, but also tend to make us feel hungrier. Such beverages have many calories, but no nutrients. They can rapidly increase blood glucose concentrations to dangerously high levels" [8].

Pomegranate wine was able to inhibit the activation of NFκB (also known as p65) in cultured bovine aortic endothelial cells (BAECs) [152, 153]. BAECs were pretreated with 4 μg/mL pomegranate wine for two hrs and then exposed to 200 units/mL of TNF-α for 5-40 mins [152]. "Treatment with TNF-α alone caused NFκB to move (translocate) to the cell nucleus within 5 mins. Treatment with pomegranate wine inhibited this migration by two to four-fold. It also reduced the binding activity of NFκB. Moreover, three compounds found in pomegranates, juice and wine (gallic acid, quercetin and catechin) also inhibited the translocation. Pretreatment with pomegranate wine did not affect the phosphorylation of IκBα or the degradation of TNF-α, but it did cause phosphorylation of serine in IκBα before the addition of TNF-α" [152]. As mentioned before, NFκB is activated by RONS, so it is important that it was not activated in these studies [152, 153]. "The NFκB system comprises a family of transcription factors that are kept inactive in the cytoplasm by a family of inhibitory proteins known as IκBs. These inhibitors are ubiquitylated and degraded by a phosphorylation reaction that is triggered by the IκB kinase (IKK)" [8]. Activation of NFκB can cause vascular disease and atherosclerosis [152]. It has a crucial role in increasing the expression of genes that code for proinflammatory "cytokines, chemokines, leukocytes, adhesion molecules, growth factors, and matrix-degrading enzymes" [152]. So, pomegranate wine inhibited the activation of NFκB by a different mechanism than red wine or N-acetylcysteine. Besides, these are all relatively safe natural substances that may be able to prevent and/or treat vascular damage that can lead to atherosclerosis [152].

The commercial POMx extract has also been shown to decrease the atherogenicity of human monocyte-derived macrophages (HMDM) in patients who have elevated cholesterol that is treated with statins in a placebo-controlled double blind study [154]. "That is, one group received simvastatin (20 mg/day) plus a vegan placebo pill, while the other group received the same dose of simvastatin plus one POMx capsule. After two months of this therapy, the group that received POMx had 26% less LDL than on day zero. The concentrations of serum thiols (including reduced glutathione) fell by 6%. They also reported that the concentration of reactive oxygen species by up to 30%, even though the assay that they used would also detect RONS. Moreover, the concentration of serum triglycerides and HMDM cholesterol decreased by 48% and 44%, respectively when compared to day zero " [154]. "These anti-atherogenic effects could reduce the risk for atherosclerosis development" [154].

Pomegranate flowers (a traditional antidiabetic medicine) or a methanolic extract of them can also improve cardiac lipid metabolism [155]. "The extract was prepared by extracting an unspecified amount of dried pomegranate flowers from India five times with five volumes (w/v) of methanol at room temperature, followed by evaporating off the methanol to produce a solid. Zucker diabetic fatty (ZDF) rats were given 500 mg/kg of this extract by oral gavage, which delivered the extract directly into their stomachs. The extract lowered the concentration of total cholesterol in the heart by about 50%. It also lowered the concentrations of total cholesterol, triglycerides and non-esterified free fatty acids in the blood plasma. Heart muscle cells (cardiac myocytes) can't make or store much of their own long chain fatty acids. They must import these from circulating blood. When the concentrations of free fatty acids are so high in cardiac myocytes that they can't be used for β-oxidation, they are incorporated into triglycerides by reactions catalyzed by acyl transferases. In metabolic syndrome and diabetes, the concentrations of free fatty acids in blood increase to dangerous levels" [155]. So, it is noteworthy that pomegranate flower extract can lower these.

The extract also decreased the expression of cardiac mRNAs that code for fatty acid transport protein, PPAR-α, carnitine palmitoyltransferase-1 (CPT-1), acyl-CoA oxidase and 5'-AMP-activated protein kinase α2, while restoring downregulated expression of mRNA coding for cardiac acetyl-CoA carboxylase [155]. "The extract and one of its bioactive chemicals, oleanolic acid, also increased the transcription of the gene coding for PPAR-α. The PPAR-α protein is a transcription factor that is activated in the heart by the non-covalent binding of long chain fatty acids and related compounds. When the concentration of cardiac PPAR-α is too high, genes coding for proteins that are involved in the uptake of free fatty acids and oxidation metabolic pathways are transcribed more. Prescription synthetic PPAR-α agonists can increase the weight of the liver, but the pomegranate flower extract did not. Fatty acid oxidation is controlled by CPT-1, so it is important that the extract decreased its expression. Moreover, fatty acid oxidation is a prominent feature of the diabetic myocardium. Finally, it is important to note that the extract did not affect the lipid metabolism of non-diabetic rats that were used as a control" [155].

High blood pressure, atherosclerosis and diabetes can also interfere with "the intricate neurovascular mechanisms underlying normal erection" [156, 157]. In an animal model of erectile dysfunction (ED), a pomegranate juice concentrate was tested by adding added 8 mL of to 992 mL of drinking water. It was given to rabbits with ED for 8 weeks. They drank an average of 3.87 mL daily of the concentrate, which was equivalent to 112 μmol of polyphenols [158]. The concentrate was prepared by crushing the whole fruit, including the peels. It "increased intracavernous blood flow, improved erectile response and smooth muscle relaxation in ED and control groups while having no significant effect on NOS expression" [158].

## ANTICANCER EFFECTS

Cancer prevention and treatment may be the best-known of all the potential health effects of pomegranates and products made from them. This has been reviewed in previous years [159-162]. As mentioned before, pomegranate peels and some of its bioactive ellagitannins can inhibit the enzyme carbonic anhydrase [134]. Since many inhibitors of this enzyme inhibit the growth of cancer cells *in vitro* and *in vivo*, this may be one way that pomegranates can help prevent cancer [159].

More recently, methanol and ethanol extracts of pomegranate leaves were found to be cytotoxic to MCF-7 breast cancer cells grown in culture [163]. They also exhibited potent antioxidant properties and inhibited the enzymes 5-lipoxygenase, acetylcholine esterase and butyrylcholine esterase [163]. Since the extractions were done at room temperature and pressure, only a portion of the bioactive substances were extracted. So, the whole leaves or an extract prepared using high temperature and pressure probably would have been more effective.

One of the earlier studies showed that the juice (fresh and fermented), an aqueous extract of the pericarps (peels) and lipophilic extracts of seed oil might be useful in preventing and, as an adjuvant, in treating breast cancer [164]. Lyophilized fresh juice (10 mg/mL) demonstrated anti-estrogenic activity when it inhibited the binding of 17-β-estradiol to its receptor *in vitro* [164]. "Phenolic compounds partly purified from the fermented juice, pericarp and seed oil inhibited key enzymes needed to biosynthesize estrogen. This includes estrogen synthase, an aromatase and 17-β- hydroxysteroid dehydrogenase Type 1. These compounds and the seed oil also inhibited the growth of MC-7 breast cancer cells grown in culture. The polyphenols in fermented juice had more of an antiproliferative effect than those extracted from fresh juice. This could be due to the hydrolysis of glycosides (sugars) from the phenols to which they were bound covalently. Some of the lipophilic compounds in the seed oil were extracted by supercritical fluid $CO_2$. They may contribute as much or more to its anti-estrogenic and anticancer properties than the phenolic compounds. This lipophilic extract contained punicic acid (as part of triglycerides), which has antioxidant activity as an inhibitor of the biosynthesis of prostaglandin (COX inhibitor). The extract also contained the phytosterols stigmasterol, campesterol, and β-sitosterol as well as γ-tocopherol and the non-steroidal phytoestrogen coumestrol. The seed oil (at 10 μg/mL) also caused a 75% inhibition of the migration of MCF-7 cells across a solubilized tissue membrane called Matrigel, suggesting that it may help to prevent metastasis. Also, the seed oil at a dose of 50 μg/mL caused 54% apoptosis in MDA-MB-435 metastatic human breast cancer cells that did not have the estrogen receptor" [164].

In the same study, fermented pomegranate juice and seed oil partially prevented the occurrence of cancer in a mouse mammary organ culture. It was an *ex vivo* model for pre-cancerous tumor inititation induced by the chemical carcinogen, 7,12-dimethyl-benz[*a*]anthracene [164]. A more recent study showed that the commercially available pomegranate extract, POMx, inhibited the proliferation of MCF-7 cells and stimulated apoptosis in them [165]. "It inhibited growth of the cells by inducing cell cycle arrest in the transition from G2 (second growth phase) to M (mitosis) followed by the induction of apoptosis. Standard antioxidants (*N*-acetyl cysteine and Trolox) did not have this effect. So, the extract acts as more than just an antioxidant. The extract also downregulated genes coding for proteins that are associated with mitosis, chromosome organization, RNA processing, DNA replication and DNA repair. It also upregulated genes coding for proteins that are involved in the regulation of apoptosis and cell proliferation. More specifically, it downregulated genes coding for proteins that are involved in repairing double strand breaks in DNA, such as *RAD50* and *BRCA1*. This downregulation was associated with increased levels of their predicted microRNAs (miRNAs). This included miR-183 (predicted target *RAD50*) and miR-24 (predicted target *BRCA1*). This suggests that the extract may regulate miRNAs that are involved in DNA repair" [165].

In a subsequent study, it was found that a lower dose (1 μg/mL) of seed oil suppressed the initiation of cancer more than a higher dose, 10 μg/mL [166]. It was suggested that this meant that "an optimal biological dose is more important and relevant than a maximally tolerated one" [159].

However, there is another possible explanation. The seed oil is primarily made up of triglycerides, which are not very soluble in DMSO and are insoluble in the aqueous growth medium that was added to the mouse mammary organ culture. The authors reported that they used dimethyl sulfoxide (DMSO) as the dose vehicle, so the seed oil was mixed with DMSO. Then this was mixed with the growth medium. Under these conditions, triglycerides will not dissolve, but will form an insoluble mass that will surround most of the bioactive phenols, preventing them from reaching the cells, especially at higher doses of seed oil. At lower doses, some of the triglycerides might form micelles that contain polyphenols instead of an insoluble mass that forms at higher doses. Such micelles could make the polyphenols more available to the cells. So, the effectiveness of the lower dose could be due to the increased exposure of the polyphenols to the mammary cells in the presence of lower concentrations of triglycerides. So, one should not jump to conclusions about the meaning of the fact that a lower dose was more effective. It should also be noted that in a study done in accordance with good laboratory practices (GLP), one must evaluate the dose formulation and stability of any dose of a test article that is to be tested [8]. That is, the dose must be shown to be homogeneous by quantitative chemical analysis. The same concentrations of the active ingredients (polyphenols for seed oil in the study under question [166] must be present in the top, middle and bottom of the dose when it is in a graduated cylinder or other suitable container. A mixture of seed oil, DMSO and aqueous growth medium will not form a stable, homogeneous mixture.

In contrast, seed oil by itself is nearly homogeneous, so it can be applied as a dermal dose. So, it is important to note that seed oil prevented the formation of skin tumors in mice when applied to the skin before applying a chemical carcinogen [167]. It also decreased the formation of tumors when given after applying the carcinogen [167]. In another study, an extract of pomegranates prevented the formation of skin tumors induced by applying a different chemical carcinogen, 12-O-tetradecanoylphorbol-13-acetate (TPA) [168]. "This was done in a multi-stage model of skin cancer in mice. The extract was prepared by squeezing the edible part (seed coat and juice) in 70% acetone plus 30% water, followed by filtration. It was analyzed by a form of mass spectrometry called MALDI-TOF MS and found to contain six anthocyanins along with oligomeric ellagitannins and hydrolyzable tannins" [168]. However, note that the mass spectrometer was tuned to make it better able to detect these phenolic compounds. Other compounds like fructose and glucose would also be expected to be present since there was 30% water in the solvent. There could also have ben some lipophilic compounds that the acetone solubilized, but these were not mentioned in the paper. Still, the extract modulated the MAPK and NF-κB signaling pathways [168]. "When the extract was applied to the skin 30 min before applying the TPA carcinogen, it inhibited the TPA-mediated increase in the activities of epidermal ornithine decarboxylase and COX-2. It also inhibited the TPA-induced phosphorylation of ERK1/2, p38 and JNK1/2, as well as the activation of NF-κB and IKKα and phosphorylation and degradation of IκBα. The authors concluded that the ability of the extract to inhibit conventional and novel biomarkers of TPA-induced carcinogenesis, indicated that it may be able to prevent other types of tumors" [168].

In a subsequent study, not just pomegranate oil, but also the juice and a commercially available extract (POMx) prevented damage to human skin cells grown in culture after being exposed to UVB light [169]. "It prevented the formation of carcinogenic cyclobutane pyrimidine dimers and 8-dihydro-2'-deoxyguanosine, as well as protein oxidation and the expression of proliferating cell nuclear antigen. They also inhibited the UVB-induced activities of the following enzymes: collagenase (MMP-1), gelatinase (MMP-2, MMP-9), stromelysin (MMP-3), marilysin (MMP-7), elastase (MMP-12) and tropoelastin. They also decreased the UVB-induced expression of the signaling protein c-fos and the phosphorylation of c-jun" [169]. In this case, the juice and POMx would probably form a homogeneous mixture with the growth medium and 0.1% DMSO, but the oil would not.

Another group found that a different pomegranate extract (from Verdure Sciences, Noblesville, IN) could protect human fibroblasts from damage caused by UVB light at doses ranging from 5 to 60 mL/L [170]. "The extract was first dissolved in DMSO and then diluted, presumably with aqueous growth medium. The extract prevented UVB light from activating NF-κB. It also downregulated the proapoptotic enzyme called caspase-3. It also decreased the UVB-induced cell cycle arrest in the S phase and partially restored the suppressed G0/G1 progression" [170].

Even though UVB does more damage, UVA can also cause benign and malignant tumors [160]. An extract of the edible portions of pomegranates was prepared by squeezing the seed coat and extracting the juice with 70% acetone plus 30% water [171]. "This extract (at 60-100 μg/mL) inhibited UVA-induced phosphorylation of signal transducer and activator of transcription 3 (STAT-3), Akt, mTOR and ERK1/2. It also inhibited the UVA-mediated increase in the expression of the proteins Ki-67 and PCNA, which are biomarkers of cell proliferation. So, this extract may be able to prevent skin cancer cells from proliferating" [171].

Pomegranate seed oil was also shown to be able to prevent colon cancer in rats that were given the chemical carcinogen azoxymethane [172]. "In this case, the oil was added to the food that was given to the rats. It increased the expression of PPAR-γ and the incorporation of conjugated linoleic acid into the triglycerides in the colon and liver of the test animals. However, the dose-response curve was not a classical one, since a lower dose had more of an effect than a higher dose" [172]. In this case, there is probably no problem with the homogeneity or bioavailability of the doses and is an example of hormesis.

In another study, pomegranate juice, ellagitannins and punicalagin all suppressed inflammatory cell signaling in human colon cancer cells grown in culture [173]. "They suppressed the expression of the tumor necrosis factor receptor (TNFR) that was induced by COX-2. Pomegranate juice also reduced the phosphorylation of the p65 protein and its binding to the NFκB response element, while eliminating the activation of Akt, which was induced by TNFα in control cells. So, pomegranate juice can help modulate inflammatory cell signaling in colon cancer cells" [173].

More recently, pomegranate polyphenolics were shown to suppress the formation of inflammation and aberrant crypt foci in rats that had colorectal cancer induced by azoxymethane [174]. "Pomegranate juice downregulated the proinflammatory enzyme nitric oxide synthase and the mRNA that codes for COX-2. It also suppressed mRNA that codes for NFκB and vascular cell adhesion molecule 1 (VCAM 1). It also inhibited the phosphorylation of Akt and the expression of mTOR. Its therapeutic effect may be due to its ability to target an inhibitory micro RNA, miR-126" [174]. This is consistent with another report that plant phytochemicals (including catechins, quercetin and ellagitannins) may be able to prevent cancer by modulating epigenetics, including miRNAs, DNA methylation and histone modifications [175]. Inhibitory micro RNAs, epigenetics and other aspects of biochemistry will be discussed further in the Appendix.

It is also quite likely that the bioactive components of pomegranate juice can act synergistically to exert their antiproliferative, apoptotic and antioxidant activities on human oral (KB, CAL27), colon (HT-29, HCT116, SW480, SW620) and prostate (RWPE-1, 22Rv1) tumor cells grown in culture [176]. The juice was more active than the individual components in inhibiting the growth of these cancer cells and in inducing apoptosis [176].

Pomegranates may also help prevent or treat liver cancer [177]. "That is, extracts of arils from four different varieties of Turkish pomegranates decreased the proliferation of human liver carcinoma cells (HepG2), grown in culture. The extracts were prepared by adding 100 mL of 80% acetone to 100 g of arils and blending the mixture in a chilled metal blender" [177]. Another study supported these results [178]. "A commercially available pomegranate emulsion (from Rimonest Ltd., Haifa, Israel) reversed the effects of the chemical carcinogen diethylnitrosamine. It inhibited the proliferation of HepG2 cells and stabilized the cell cycle. It also induced apoptosis by upregulating the expression of the proapoptotic protein Bax and downregulating the antiapoptotic protein Bcl-2. The emulsion also reduced the expression of hepatic $\beta$-catenin and augmented glycogen synthase kinase-3β" [178]. They claimed to have analyzed the emulsion which was described in another paper as a "proprietary combination of pomegranate aqueous phase extract and pomegranate seed oil" [179]. However, this paper just cited another article [180] that just analyzed some seed oil by LC-MS, and not GC. So, when they [178-180] claimed that it contained primarily "linoleic acid (~20%), $\alpha$-linolenic acid (~3%), and some conjugated linolenic acids like PA (9Z, 11*E*, 13Z, ~65%) and AEA (9Z, 11*E*, 13*E*, ~7%) [180], they were at least partly wrong, because the seed oil is almost all triglycerides, but with a smaller percentage being plant sterols. If they truly used LC-MS without first hydrolyzing the triglycerides, all they would have seen would have been genuine free fatty acids, which would be present only at very low levels. Instead, the so-called fatty acid profile that they reported closely resembles the known fatty acyl composition of pomegranate seed oil triglycerides, which would never have been solubilized by the methanol that they reportedly used. So, it is most likely that the fatty acyls in the triglycerides exerted their effects on the cells only after being hydrolyzed in reactions catalyzed by lipases.

Pomegranates may also be useful in helping to treat urinary bladder and urothelial cancer [181]. An ethanol extract of fresh pomegranate juice restricted the proliferation of human urinary bladder urothelial carcinoma cells grown in culture [181]. "It arrested the progression of the cell cycle in the S (synthesis) phase by increasing the concentration of cyclin A and decreasing the expression of cyclin-dependent kinase 1. It also activated pro-caspases 3, 8 and 9, while increasing the ratio of Bax to Bcl-2. So, its activity could be based on triggering apoptosis through death receptor and mitochondrial damage pathways" [181].

Some people also study the anticancer effects of individual components of pomegranates. This is especially true for ellagic acid, which was reviewed in 2008 [182]. They cited one of the earliest papers that described how daily consumption of pomegranate juice lowered the oxidation of LDL cholesterol, leading to the elimination of plaques in coronary arteries [183]. "The beneficial effects of pomegranate juice on atherosclerotic disease in animal models have since been ascribed to the presence of ellagic acid" [182].

Despite all of these *in vitro* studies, it is important to demonstrate the safety of pomegranates. As mentioned previously, this may not be necessary, since the fruit, seed oil, juice and extracts prepared from them are generally regarded as safe (GRAS) [6]. However, the root and stem barks are toxic, due to the presence of alkaloids [7]. On the other hand, high doses of ellagitannins were found to be safe when given to rats [7]. This was reinforced by some studies done on healthy human subjects [9-11]. Still, it is important to know if pomegranates are safe when consumed by a more vulnerable portion of the human population, such as those who have metabolic syndrome (obesity). So, one group looked at the safety and antioxidant activity of the extract POMx on overweight people with a large waist size [184]. "One group consumed a placebo while the other two groups ate one or two POMx pills per day, which provided 435 and 870 mg gallic acid equivalents, respectively. The blood serum of people in the groups receiving POMx had a higher antioxidant capacity than the controls who received the placebo. The groups that received the POMx had no adverse effects" [184].

After establishing safety, the next step in getting an investigational new drug considered for approval by the FDA or other governments' regulatory agencies is to do clinical studies. Some of these have been done on the most widely studied form of cancer that may be prevented and/or treated by pomegranates. That is prostate cancer. Moreover, pomegranates and products made from them may be good for men's overall urological health [185]. Their antioxidants may help with chronic inflammation, which can lead to obstructive and irritative lower urinary tract symptoms (LUTS) of benign prostate hyperplasia (BPH), erectile dysfunction (ED) and prostate cancer [185]. For example, the commercially available POMXL liquid extract (see page 35) was able to treat ED in an animal model (rabbits) in which ED was caused by atherosclerosis [186], which can also help cause ED in men [185]. The rabbits were divided into four groups. One received placebo (tap water) and the other three groups received enough POMXL to get 30, 60 or 120 mg polyphenols per day per 3.5 kg rabbit [186]. "The erectile tissues of control rabbits had impaired endothelium-dependent relaxation, diffused fibrosis, increased oxidative products, upregulated expression of genes coding for superoxide dismutase (SOD) and aldose reductase (AR), mitochondrial and endothelial structural damage and increased caveolae. The upregulated antioxidant enzymes, SOD and AR, were not active enough to prevent ischemic erectile tissue from suffering oxidative injury. The groups receiving POMXL had improved intracavernosal blood flow, erectile activity, smooth muscle relaxation and less fibrosis than the group receiving placebo. It also decreased the concentration of oxidative products, thus protecting the mitochondrial, endothelial and caveolae structural integrity" [186]. It should be noted that this extract is quite different from solvent extracts that were used in many of the studies described earlier in this chapter. The only solvent that was used was an aqueous solution containing enzymes. No organic solvents were used. It is possible that much higher levels of polyphenols could be extracted if pressurized water or ethanol were used to re-extract the residue remaining after the initial treatment with the enzymes. Moreover, many of the polyphenols in POMXL are also in the juice and whole pomegranates. So, these may also be effective in helping to prevent or treat ED.

A similar study found increased intracavernosal blood flow, improved erectile response, and smooth muscle relaxation [187]. They also found that long term consumption of POMXL helped to prevent erectile tissue fibrosis [187].

The treatment of prostate cancer may also aided by pomegranates and products made from them [185]. That is, seed oil, juice, fermented juice and compounds isolated from pomegranates may be effective in preventing and/or treating prostate cancer [160]. In 2004 there was a report describing a multicenter study of the effects of oil obtained two ways: by cold-pressing the pomegranate seeds and by extracting them with supercritical fluid (SFE) $CO_2$. Fermented juice polyphenols and pericarp polyphenols were studied as well [188]. These studies examined human prostate cancer cell xenograft growth *in vivo,* and/or proliferation, cell cycle distribution, apoptosis, gene expression, and invasion across Matrigel, *in vitro* [188]. "Both preparations of seed oil, the fermented juice polyphenols and pericarp polyphenols all inhibited the proliferation of LNCaP, PC-3, and DU 145 human prostate cancer cell lines *in vitro*. The LNCaP cells have androgen receptors and secrete prostate-specific antigen (PSA), but are not invasive. The PC-3 and DU145cell lines do not have androgen receptors or secrete PSA. The PC-3 cells are invasive and have a high metastatic potential, but the DU 145 cell line does not. The dose of pericarp polyphenols needed to inhibit cell proliferation of the prostate cancer cell line LNCaP by 50% ($ED_{50}$) was 70 μg/mL" [188]. They were isolated from an aqueous extract of pericarps [164].

That is, "the pericarps (peels) were separated from the seeds after juicing and placed in a large stainless steel pot and covered with water [164]. "The pot was heated over an open fire and allowed to cook at a rolling boil for 45 min. The liquor was then decanted and further concentrated over the fire until it was 50% solids. This was combined with twice its volume of ethyl acetate, left for 8 hr, and dried in a vacuum evaporator at 40 °C to obtain a sticky residue" [164]. It is quite likely that the extensive boiling hydrolyzed off sugars that were attached to some of the phenolic compounds, especially anthocyanins. Also, it is not clear why ethyl acetate was added and then evaporated off. Some of it would have mixed with the water, but most of it would have formed a separate phase. If the ethyl acetate phase had been separated from the water before evaporating it off, the solids obtained would have been enriched in phenolic compounds and would not have contained any sugars. If it was not separated before evaporating off the water plus ethyl acetate, there would have still been phenolic compounds in it, but also water-soluble sugars.

Either way, when using the "pericarp polyphenols", it took a higher dose (250 μg/mL) to affect normal prostate epithelial cells (hPrEC) [188]. "The effects on the cancer cells were mediated by changes in both the distribution of cells in different parts of the cell cycle and the induction of apoptosis. At doses of 30, 45 and 35 μg/mL of the pericarp polyphenols, fermented juice polyphenols, and seed oil, respectively, the androgen-independent cell line DU 145 was shifted towards the G2/M phase of the cell cycle and there was a modest induction of apoptosis. In the other two cell lines, the apoptotic response predominated. Apoptosis in the PC-3 cells treated with pericarp polyphenols occurred at least partially through a caspase 3-mediated pathway. These cellular effects coincided with rapid changes in the levels of mRNA expressed by gene targets. The 4-hour treatment of DU 145 cells with 35 μg/mL seed oil up-regulated the cyclin-dependent kinase inhibitor p21*(waf1/cip1)* and down-regulated c-myc. At the same time, all the pomegranate preparations tested suppressed the migration of PC-3 cells through Matrigel, which is a measure of the invasiveness of the cells" [188]. This is important because prostate cancer by itself is seldom fatal, but can become fatal if the cancer invades other organs in the process of metastasis [8]. Finally, pericarp polyphenols and SFE-extracted seed oil inhibited the growth of PC-3 xenografts in athymic nude mice [188]. Such mice are genetically engineered in a way that they have either a deteriorated or non-existent thymus gland that is needed to biosynthesize immune cells [8]. This enables them to receive xenografts without rejecting them.

In another study, a well-described extract of pomegranate juice inhibited the growth of highly aggressive human prostate cancer PC3 cells, while stimulating apoptosis [189]. "To prepare the extract, they peeled fresh pomegranates. The edible portions were squeezed in 70% acetone plus 30% water at a ratio of 1:20 (w/v). The extract was filtered, condensed and freeze-dried to obtain a product that contained six anthocyanins (pelargonidin 3-glucoside, cyanidin 3-glucoside, delphinidin 3-glucoside, pelargonidin 3,5-diglucoside, cyanidin 3,5-diglucoside, and delphinidin 3,5-diglucoside), along with various ellagitannins and hydrolyzable tannins, as shown by MALDI-TOF MS analysis. The extract was first dissolved in DMSO, then diluted into the growth medium so that the concentration of DMSO was 0.1% (v/v). This very likely produced a reproducibly homogeneous dose. The diluted extract induced the transcription of genes coding for the proapoptotic proteins Bax and Bak, while downregulating the antiapoptotic proteins Bcl-$X_L$ and Bcl-2. It also affected progression through the cell cycle" [189].

Passage through the cell cycle is controlled by a family of protein kinase complexes [189, 190]. "Each multi-protein complex contains a cyclin-dependent kinase (cdk) as the catalytic protein and an activating protein, called a cyclin. Cyclins D and E are active during G1–S (first growth, G1 and DNA synthesis, S) phase of the cell cycle. In controlled cell growth, the non-covalent binding of cyclins D and E with cdk2, cdk4, or cdk6 induces the phosphorylation of the tumor suppressor called retinoblastoma (Rb). This causes it to be released from elongation factor 2 (E2F), resulting in progression of the cell cycle and cellular proliferation. Any defect in this process can change the regulation of the cell cycle, causing uncontrolled proliferation and cancer. During the progression of the cell cycle, the cdk–cyclin complexes are inhibited by binding to cyclin kinase inhibitors such as the WAF1 and KIP1 families of proteins" [189, 190].

So, the effects of the extract of pomegranate juice on regulatory proteins that are active in the first growth ($G_1$) phase of the cell cycle were determined [189]. "The proteins WAF1 (also known as p21 and cyclin-dependent kinase inhibitor 1) and KIP1 (also known as cyclin-dependent kinase inhibitor 1B and p27) were upregulated during the $G_1$ phase, while apoptosis was stimulated. Moreover, the extract down-regulated the regulatory proteins that are active in $G_1$. This included cyclins D1, D2, and E and cdk2, cdk4, and cdk6. Proteins in the B-cell lymphoma 2 (BCl-2) family also help regulate apoptosis. Bcl-2 suppresses apoptosis and is found at relatively high concentrations in over half of all human tumors. It can be controlled by forming a complex with the proapoptotic protein Bax. So, the ratio of Bax/Bcl plays a crucial role in deciding if a cell will die or survive. Thus, it is important that the extract decreased the expression of Bcl and increased Bax. Apoptosis is essential for proper tissue homeostasis and removal of unwanted cells. In conclusion, the extract inhibited the growth of cancer cells by arresting progress in the cell cycle and inducing apoptosis" [189].

This same extract also inhibited the growth of androgen-sensitive CWR22Rv1 prostate cancer cells that were implanted into athymic nude mice, which showed that this can also occur *in vivo* [189]. These cells form rapid and reproducible tumors in nude mice and secrete significant amounts of PSA in the bloodstream of the host" [189]. "One group of these mice (control) drank tap water, while two other groups drank tap water that had either 0.1 or 0.2% pomegranate extract. The control group developed tumors with an average volume of 1300 $mm^3$ after an average of 31 days. The group receiving 0.1% extract developed this size of tumors in an average of 39 days. The group receiving 0.2% extract developed this after 47 days. In addition, the concentrations of PSA in the blood serum were $7.9 \pm 0.82$, $2.5 \pm 0.58$, and $1.2 \pm 0.97$ ng/mL in controls, 0.1% extract-fed, and 0.2% extract-fed mice, respectively" [189]. Even though PSA is not a good predictor of developing prostate cancer [8], it is used to follow the progression of prostate cancer [189].

As mentioned previously, prostate cancer, like most cancers, becomes fatal only after the cancer moves to other organs in a process called metastasis [8]. A crucial step in metastasis is the formation of new blood vessels in a process called angiogenesis. So, it is important that the ellagitannin-rich extract POMx inhibited angiogenesis in prostate cancer both *in vitro* and *in vivo* [191]. "The extract was dissolved in phosphate buffered saline solution and added to the cell culture medium used to grow human prostate cancer cells (LNCaP) that were dependent on androgens. Cells were grown in normoxic and hypoxic conditions. POMx inhibited the proliferation of these cells under both conditions. Moreover, it reduced the concentration of vascular endothelial growth factor (VEGF) and the expression of heat inducible factor 1α (HIF-1α) under hypoxic conditions. HIF-1α induces the expression of genes coding for proteins needed for angiogenesis under hypoxic conditions. VEGF helps regulate angiogenesis. Its concentration is elevated in prostate cancer. Androgens (male hormones) activate the expression of the genes coding for HIF-1α and VEGF. These proteins were also expressed at lower levels in mice that had severe combined immunodeficiency (SCID) and received injections of androgen-dependent LAPC4 prostate cancer cells in the *in vivo* study" [191]. Since prostate cancer grows very slowly, POMx may help save the lives of many men just by slowing it down even more.

"Prostate cancer is dependent on circulating testosterone in its early stages and is treatable with radiation and surgery" [192]. "However, recurrent prostate tumors advance to an androgen-independent state in which they progress in the absence of circulating testosterone, leading to metastasis and death" [192]. "During the development of androgen independence, prostate cancer cells increase the intracellular synthesis of testosterone, which maintains cancer cell growth in the absence of testosterone circulating in the blood. Overexpression of the androgen receptor (AR) occurs in androgen-independent prostate cancer. It may help promote the development of androgen independence. To study this, a cell line called LNCaP–AR can be used. It is genetically engineered to overexpress AR, but is otherwise similar to the widely studied LNCaP cell line. In one study, the effects of pomegranate juice, POMx, punicalagin and ellagic acid on the expression of genes coding for key androgen-synthesizing enzymes and the AR were determined. This included HSD3B2 (3β-hydroxysteroid dehydrogenase type 2), AKR1C3 (aldo-keto reductase family 1 member C3) and SRD5A1 (steroid 5α reductase type 1). These pomegranate products inhibited the expression of these genes and the AR most consistently in the LNCaP–AR cell line. Also, the AR may be of particular importance in androgen-independent prostate cancer cells and the type of human prostate cancers in which the AR is up-regulated. The POMx contained ellagitannins (37-40%) and 3.4% ellagic acid, but no anthocyanins. Pomegranate juice (POM Wonderful) was used as a concentrate powder. It contained punicalagin, ellagic acid and anthocyanins. POMx, juice, punicalagin and ellagic acid all inhibited cell proliferation in a dose-dependent manner in the androgen-dependent cell line (LNCaP) as well as the androgen-independent LNCaP–AR and DU-145 human prostate cancer cell lines. They also induced apoptosis" [192].

In a recent study, a different type of pomegranate extract was able to reduce the biosynthesis of androgens (testosterone and dihydrotestosterone (DHT), dehydroepiandrosterone (DHEA), androstenedione, androsterone, and pregnenolone) at doses of 2 – 12 μg/mL in two prostate cancer cell lines, LNCaP and 22RV1 [193]. At a dose of 0.17 g/L in drinking water, the extract also reduced the serum PSA concentration and biosynthesis of these androgens in a mouse model of prostate cancer produced by removing (knocking out) the signaling protein PTEN [193]. PTEN is also called phosphatase and tensin homolog [8]. It is a tumor suppressor that catalyzes the opposite reaction that is catalyzed by the oncoprotein PI3K [8]. To prepare the extract, whole pomegranate fruit was broken into small pieces and extracted using 100% ethanol [193]. The ethanol solution was evaporated off [193]. "The dried powder was re-dissolved in ethanol, and then centrifuged to separate the ethanol-soluble fraction from the water-soluble residue. The residue was dried and reconstituted in an ethanol–water (70:30) mixture to form a solution of 200 mg/mL, which was used to prepare the cell treatment media used in the experiment" [193].

These results are important because "androgen deprivation therapy is the primary treatment for patients with metastatic prostate cancer" [193, 194]. Even though this is effective at first, within 18-20 months most patients progress to a deadly form of cancer that is resistant to castration [193, 194]. The enzymes that catalyze the biosynthesis of androgens are expressed at higher levels in castration-resistant metastases, so they may be good therapeutic targets [193, 194]. By inhibiting them, the pomegranate extract and possibly even the juice may be good supplements to help treat castration-resistant prostate cancer [193].

Not only pomegranate extracts, but also individual compounds and their metabolites may work synergistically to help prevent and/or treat prostate cancer [195-197]. For example, the individual *in vitro* antiproliferative, apoptotic and antioxidant activities of punicalagin, ellagic acid and a total pomegranate tannin extract were enhanced when they combined with other components in pomegranate juice [195]. "The antiproliferative activities of punicalagin, ellagic acid and a total pomegranate tannin extract were evaluated on human oral (KB, CAL27), colon (HT-29, HCT116, SW480, SW620) and prostate (RWPE-1, 22Rv1) tumor cells grown in cell culture. The total pomegranate tannins were extracted from the peel (or husks) and analyzed for purity by HPLC and LC-MS. They consisted of 85% punicalagin isomers (all with molecular weights of 1084) and about 12% other ellagitannins and ellagic acid glycosides. They were dissolved in DMSO and then diluted with growth medium to obtain the desired concentrations. Punicalagin, EA and TPT were tested at concentrations of 12.5, 25, 50 and 100 μg/mL for 48 hr by adding them to the cells. Pomegranate juice concentrate from the 'Wonderful' cultivar was tested at a concentration that delivered equivalent amounts of punicalagin to see if there were synergistic effects. It contained 1.74 and 0.14 mg/mL of punicalagin and ellagic acid, respectively. The antioxidant effects were measured by the Trolox equivalent antioxidant capacity (TEAC) and inhibition of lipid peroxidation assays induced by $Fe^{2+}$. The pomegranate juice had the highest antioxidant and antiproliferative activities against the cell lines. Since cancer cells exist in a state of oxidative stress, the authors felt that the antioxidant activities of the test articles were responsible for their anticancer activities" [195]. Finally, this paper [195] was important because it showed the synergistic activity of the juice. In some ways, it was ahead of its time. As mentioned in the Introduction of this book, until recently, very few peer-reviewed scientific journals accepted articles that described the biochemical activities or health effects of a mixture of compounds or a whole food or juice. Such studies were often considered to be "dirty" and difficult to interpret. Fortunately, this trend is disappearing as medicinal chemistry is becoming a fusion of traditional and western medicine that accepts the idea of synergism [8].

A similar study showed that the effects of seed oil and phenolic compounds isolated from fermented juice and peels were synergistic on the antiproliferative, metstatic potential and ability to inhibit a key enzyme, secretory phospholipase 2 (sPLA2) [196]. The sPLA2 enzyme catalyzes the hydrolysis of the fatty acyl on carbon number 2 of phospholipids [196]. Arachidonic acid is often attached to this carbon, so it is produced when the arachidonoyl fatty acyl is hydrolyzed. Arachidonic acid is a precursor of proinflammatory eicosanoids. So, sPLA2 is also proinflammatory and materials or compounds that inhibit it are anti-inflammatory. The oil was dissolved in hot ethanol to give a stock solution that supposedly had a concentration of 8.25 μg/mL. It should be noted that the solubility of triglycerides in ethanol at room temperature is very low and they are almost completely insoluble in the aqueous growth medium used in the cell cultures.

The authors cited two papers that were described above [164, 183] to tell how the phenolic compounds were isolated. Although one of them was unclear [164], the other [183] did clearly state that fermented pomegranate juice was extracted with two times its volume of ethyl acetate, which extracted the phenolic compounds from the aqueous fermented juice. The ethyl acetate was evaporated off to provide the phenolic compounds [183]. The main difference between the fermented juice and fresh juice is that the process of fermentation removes sugars from phenolic compounds to which they are covalently attached. It then converts the sugars to ethanol. It should be noted that sugars can also be removed from such compounds in the stomach by the process of digestion.

So, the effects of oil and phenolic compounds isolated from the fermented juice and peels on the proliferation of DU 145 and PC-3 human prostate cancer cells were determined [196]. When a dose of oil that was supposedly 16.5 μg/mL was given to the cells, it had no effect on their proliferation (growth). Since this is higher than the supposed concentration of the stock solution of oil in ethanol, it is quite likely that there was a mistake or mis-print. The concentration of the stock solution in ethanol was probably closer to 6.25 mg/mL (not μg/mL). This points out an important fact. Articles in peer reviewed journals can have mistakes, especially in the brief descriptions of how doses are prepared. Moreover, no attempt was made to test the dose (16.5 μg/mL) to see if it was homogeneous. It almost certainly was not, since triglycerides are not soluble in aqueous growth media. This would not have been permitted in a study done in accordance with good laboratory practices (GLP), which is required for any investigational new drug that is being tested.

The paper in question [196] also gave relatively low doses of 0 – 100 μg/mL of the fermented juice phenols. They did not affect the growth of DU 145 cells [196]. However, when the oil and fermented juice or peel phenols were given together, the growth of the DU 145 cells was inhibited [196]. The invasiveness of PC-3 cells and activity of sPLA2 showed similar effects. The authors claimed that this meant that synergism had been demonstrated [196]. Although this is possible, it is just as likely that the non-homogeneous dose of the oil delivered very different concentrations of active ingredients to the DU 145 cells in the three different experiments (oil by itself, oil plus fermented juice phenols and oil plus peel phenols).

So, clinical studies [197] that use the whole juice are often more informative than studies done on isolated fractions or oil that is mixed with aqueous growth media [196]. That is, pomegranate juice ('Wonderful' cultivar) was given to men whose concentrations of PSA were rising after being treated with surgery or radiation [197]. The decision on the appropriate doses was made based on the effects of 0 to 15 ounces (about 0 to 420 g) on the "antioxidant effect" of them on healthy male subjects [197]. Although the paper in question [197] said that the results were unpublished, it is most likely that the antioxidant capacity of the blood was measured. "In any case, the optimal dose of juice was 6 to 8 ounces per day for 13 months. No serious adverse effects were reported by the subjects of the trial. The mean doubling time for PSA was increased from 15 to 54 months. Analyses of the subjects' blood serum showed a 12% decrease in cell proliferation, a 17% increase in apoptosis, a 23% increase in nitric oxide and significant reductions in oxidative state and sensitivity to oxidation of lipids [197].

Based on the results of the clinical trial [197], a subsequent study on severe combined immunodeficiency (SCID) mice looked at an extract of juice and metabolites (urolithins) of ellagitannins, the most abundant polyphenols in the juice [198]. "The mice received xenograft implants of androgen-dependent LAPC-4 prostate cancer cells (200000 cells per animal). First, an extract of the juice was prepared. It was enriched in ellagitannins. It had 37% ellagitannins as punicalagin equivalents. The dose given to mice (0.8 mg ellagitannins) was calculated based on ten times the equivalent dose that would be administered to a 70 kg human who would consume 240 mL of juice that was not concentrated. In another experiment, the metabolite urolithin A was given to mice at a dose of 0.3 mg per animal. Both the juice extract and urolithin A were suspended in 50 μL of phosphate buffered saline or 10% glucose in water for intraperitoneal or oral administration, respectively. The prostate cancer cell growth was inhibited in the mice that received pomegranate juice. Tumor tissues were analyzed for metabolites. Ellagic acid was present at higher concentrations in tumors collected from the mice who received the juice, which demonstrated the important fact that at least one of the bioactive compounds reached the tumors. In normal, healthy (wild-type) mice, ellagic acid was detected in the blood plasma at 0.5 hr and was cleared after 2 hr. Intraperitoneal doses produced higher concentrations than oral doses in not just blood serum, but also prostate glands, intestines, colons, and livers of the mice. Finally, ellagic acid, urolithin A and urolithin A derivatives produced in the lab by cultured gut bacteria (microflora) were given to androgen-dependent (LNCaP) and independent CaP cell lines (LNCaP-AR, DU145, and 22RV1). These ellagitannin metabolites were concentrated in the prostate glands of the mice. The metabolites had dose-dependent antiproliferative effects on all of the cell lines. The methylated urolithin A metabolite had the lowest $IC_{50}$ (highest bioactivity) of all the metabolites. The authors concluded that 'Wonderful' pomegranate juice may be helpful in preventing prostate cancer in mice" [198]. So, it may be reasonable to conclude that the juice may also help prevent prostate cancer in men.

## Effects on Neurodegenerative Diseases

Smoldering inflammation can also lead to neurodegenerative diseases [8]. So, the antioxidants in pomegranates may be able to prevent some of them. For example, some patients suffer from memory loss after heart surgery [199]. "It can be caused by a lack of blood flow (ischemia) caused by small blood clots (microemboli) that become dislodged during the surgery. Five patients were given two POMx capsules – one in the morning and two in the evening. This began two weeks before surgery and continued six weeks after surgery. Each pill contained about 375 mg punicalagins, 93 mg of anthocyanins, 29 mg of ellagic acid and 100 mg of other tannins. There was also a group of six people who received a placebo. Both groups were given memory tests before and six weeks after surgery. This included the WAIS-III Digit Span subtest, Hopkins Verbal Learning Test-Revised (HVLT-R), the Rey Complex Figure Test with recognition trial (RCFT), the Logical Memory subtest of Wechsler Memory Scales (WMS-III), and the Brief Visuospatial Memory Test-Revised (BVMT-R). The WAIS-III Digit Span subtest tested working memory. The HVLT-R and MS-III Logical Memory tests measured verbal learning and memory. The RCFT and BVMT-R tests evaluated visuospatial memory. The patients who received placebo suffered deficits in memory retention, while those who received the POMx pills were not only protected against this, but also actually improved their memory retention performance for up to six weeks after surgery when compared to their memory before surgery" [199].

In another study, pomegranate juice was helpful in improving the behavior of mice that had an animal model of Alzheimer's disease [200]. One of the features of Alzheimer's disease is the development of a soluble amyloid β peptide containing 42 amino acids (Aβ42) and amyloid plaques in the hippocampus [8], which will be discussed further in the Appendix. The mouse model consisted of transgenic mice (APPsw/Tg2576) [200]. "Beginning at six months of age, one group received pomegranate juice from the 'Wonderful' cultivar that was diluted 1:40 [200]. "The juice contained 84% water, 14% carbohydrates, 0.48% ash, 0.4% citric acid, 0.1% protein, 0.02% fat and 1% other, including polyphenols (phenolic acids and flavonoids). Phenolic acids included 115 μg/mL ellagic acid and 5 μg/mL gallic acid. Flavonoids included 1880 μg/mL hydrolyzable tannins (that is, gallotannins, ellagitannins, punicalagin) and 369 μg/mL anthocyanins and their glycosides (e.g., cyanidin, delphinidin, pelargonidin). The placebo group just received sugar water from 6 to 12.5 months of age. The mice that received pomegranate juice learned water maze tasks more quickly and swam faster than controls. They also had significantly less (~50%) accumulation of soluble Aβ42 and amyloid deposition in the hippocampus as compared to control mice" [200].

In a more recent study, pomegranate polyphenols and a POMx inhibited the activation of nuclear factor of activated T-cell activity (NFAT) and microglia in a mouse model of Alzheimer's disease [201].

Pomegranates may also help with depression [202]. "A very interesting recent study used pomegranate juice that was prepared from the seeds and the juice-filled arils surrounding them. It was made from an Iranian cultivar from the agricultural garden in Shiraz. The seeds of the fruits containing intact arils or sacs were separated from the pericarp. They were ground in a juicer so that the seeds broke. Fifty male BALB/c mice, weighing 25 - 30 g were divided into five groups. As a negative control, one group was just given normal saline (0.9% NaCl in water). The tricyclic prescription antidepressant fluoxetine (Prozac®) was used as a positive control in another group. It was given as an intraperitoneal dose of 10 mg/kg in normal saline 1, 12 and 24 hr before putting the mice through a stressful swimming test. The other three groups received doses of 1, 10 and 20 mL/kg pomegranate juice as a gavage also 1, 12 and 24 hr before the stressful test. The mice were "forced to swim in a transparent glass vessel (25 cm high, 15 cm in diameter) filled with (12.5 cm high) water at 21°C - 24°C. The total duration of immobility (in seconds) was measured during the last 4 min of a single 6 min test session" [202]. "When mice became depressed they became nearly immobile, making no further attempts to escape except the movements necessary to keep their heads above the water. The time in which they were immobile was shortened in the positive control (fluoxetine) and in mice given an acute dose of 20 mL/kg of the juice. In addition, when mice were given the juice (1 or 10 mL/kg) once daily for seven days before the stress test, the immobility time decreased. This also happened when mice were given the juice repeatedly (1, 12 and 24 hr before the stress test) over a 24 hr period. In conclusion, this preparation of pomegranate juice exhibited antidepressant effects when given once or three times to mice. This is consistent with the antidepressant activity of pomegranate juice described in Iranian folk medicine" [202]. It is also important to note that the acceptance of this article in a peer reviewed journal is a good sign that scientists are accepting the fact that whole foods and juices should be tested. That is, no solvent extract or individual compound was evaluated. Even though the title of the article [202] used the term extract, no solvents were used. The whole arils and seeds were used and they were separated from the peel (pericarp). So, many people would not consider this an extract, but a whole food. It contained not only phenolic compounds, but also triglycerides from the seeds that have a high concentration of unsaturated fatty acyls, especially punicoyl or punicic acid. It also had the phytosterols stigmasterol, campesterol, and β-sitosterol as well as γ-tocopherol and the non-steroidal phytoestrogen coumestrol. The bioactive components of this whole food may have acted synergistically to exert their health effects. Hopefully, these results will encourage manufacturers of pomegranate juice to include the seeds in the preparation.

Pomegranates have also been shown to improve affective and motor behavior in mice that were exposed to radiation [203]. "This study was done because Astronauts and Cosmonauts are exposed to ionizing radiation when they are in lower Earth orbit and some day may go on a very long trip to Mars. So, two groups of four month old male and female C57BL/6 mice were exposed to a whole body radiation beam of proton radiation (150MeV/n) at a dose of 2Gy for 5 minutes. One group drank 'Wonderful' pomegranate juice that was diluted 1:20 with water. This provided about 0.6 mg of polyphenols and was the equivalent of an adult human drinking two 8-ounce glasses of pomegranate juice every day. Another group drank sugar water that had the same amount of sugar that is in the diluted pomegranate juice. As a negative control, another group of the same type of mice were kept in the same lab environment but were not exposed to this radiation. Eight weeks later, they were subjected to tests of cognitive and motor functions over a two-week time span. The group that was irradiated and drank sugar water did worse on the tests for depression, overall activity and sensorimotor coordination and balance than the mice that drank pomegranate juice and were irradiated. There were some important differences between males and females. Females exposed to radiation had greater anxiety-like behaviors than males. Drinking pomegranate juice blocked this effect of radiation on female behavior. Pomegranate juice treatment led to gender-specific behavioral changes even in the absence of radiation-induced deficits. Although radiation had no detectable effects on sensorimotor coordination and balance, females performed significantly better than males on the test for sensorimotor coordination and balance. The authors pointed out that dietary polyphenols can stimulate neurogenesis in the hippocampus. So, the polyphenols in pomegranate juice may protect against radiation damage to the hippocampus by stimulating the growth of new neurons in the hippocampus" [203].

However, pomegranates may not be good for people who have Parkinson's disease [204]. "In a recent study, researchers tested pomegranate juice to see if it could help relieve oxidative stress and degeneration in the dopamine neurons in the substantia nigra. They used an animal model of Parkinson's disease in which rats were given rotenone, which causes a syndrome similar to Parkinson's disease by causing oxidative damage and inflammation in the substantia nigra of the brain. Rats were divided into four groups. The first group received a dose vehicle consisting of Miglyol 812N. The second group received Miglyol 812N plus pomegranate juice. The third group received Miglyol 812N plus rotenone and the fourth group was given Miglyol 812N plus rotenone plus 'Wonderful' pomegranate juice (diluted 1:40 in water)" [204]. The authors claimed that the dose vehicle was "a medium chain fatty acid" [204], even though the Material Safety Data Sheet (MSDS) clearly states that it is a "Mixture of decanoyl- and octanoyl glycerides (esters of fatty acids plus glycerol)" [205]. The postural stability of the rats was measured, since instability is a hallmark of Parkinson's disease. Neurochemical and immunoblotting analyses were done on the striatum and substantia nigra. That is, the concentrations of dopamine and its metabolites (3,4-dihydroxyphenylacetic acid and homovanillic acid) were measured. The amounts of dopamine neurons in the striatum and substantia nigra were determined by measuring the amount of the enzyme tyrosine hydroxylase, which catalyzes the conversion of L-tyrosine to L-3,4-dihydroxyphenylalanine (L-DOPA). Nigral sections were evaluated for oxidative and nitrosative stress (stress caused by reactive nitrogen substances). That is, the concentrations of nitrotyrosine and 4-hydroxynonenal were determined. Neuroinflammatory responses were analyzed in the nigral dopaminergic neurons by quantifying the activities of inducible nitric oxide synthase (iNOS), caspase and $Ca^{2+}$-binding adaptor molecule 1 (a marker for microglial cells). Finally, the rats were tested for rearing behavior. The rats who received rotenone exhibited fewer rears, had less dopamine, more active iNOS and caspase, oxidative damage and inflammation in the substantia nigra while suffering a severe loss of dopamine fibers and neurons. Also, their microglia were activated, unlike the ones who just received the dose vehicle. Unfortunately, the rats who received rotenone and pomegranate juice had even more of these symptoms of parkinsonism" [204]. Even though the authors did not understand the important differences between triglycerides and fatty acids, they did understand neurology. They were in the department of neurology in an institute for neurological diseases. So, it is quite likely that the conclusions of the study were valid. That is, pomegranate juice probably did exacerbate the damage caused by this animal model of Parkinson's disease. It is completely unknown whether this is also true for humans who have Parkinson's disease, but it might be. It is difficult to imagine how an ethical study could be done on humans to see if pomegranate juice makes the symptoms worse and/or accelerates the progression of the disease. Still, it might be best to err on the side of caution and recommend that people who have Parkinson's disease refrain from drinking pomegranate juice.

Finally, it should be added that this is not the first case of a fruit causing Parkinson's disease. At least two fruits in the Annonaceae family, graviola (*Annona muricata*) and the North American pawpaw (*A. triloba*) may cause tau pathologies [206-212]. Overconsumption of graviola caused an atypical form of Parkinson's disease on the Caribbean island of Guadeloupe [206-212], some areas of London, and the Pacific islands of New Caledonia and Guam [208]. Also, ethyl acetate extracts of pawpaw were found to be toxic in primary rat cortical neurons [210]. Both graviola and the pawpaw contain the neurotoxin, annonacin [210, 213, 214]. Annonacin and another neurotoxin, squamocin, are acetogenins (ACGs). They are derivatives of fatty acids and have 32 or 34 carbons, including a long chain alkyl group on one end and a monounsaturated or unsaturated lactone on the other end [215]. It is completely unknown whether pomegranate juice contains any such neurotoxins, though, so its potential neurotoxicity remains an open question at this time.

## ANTIBACTERIAL AND ANTIVIRAL EFFECTS

There has been paper published on the antibacterial, anti-inflammatory and anti-allergic effects of an extract made from Chinese pomegranate rinds (pericarps) [216]. "The peels were dried and ground into a powder. About 0.5 kg of this was extracted twice with 90% methanol containing 10% water by refluxing the mixture for 1 hr. The solvent was evaporated off in a vaccum (*in vacuo*). The residue obtained was suspended in 2% aqueous acetic acid and partitioned with ethyl acetate. The pooled ethyl acetate fractions were then evaporated to dryness *in vacuo*. The ability of this extract to inhibit the growth of bacteria and kill them was measured. That is, the minimum inhibitory concentration (MIC) on growth and minimal bactericidal concentration (MBC) that killed *Propionibacterium acnes* (DMST 14916), *Salmonella typhimurium* (DMST 562), *Shigella sonnei* (DMST 561), *Staphylococcus aureus* (ATCC 25923), *S. aureus* (isolated strains PSU01-PSU07), *Staphylococcus epidermidis* (ATCC 14990), *Salmonella typhi*, and *Eschericia coli* (ATCC 25922) were determined. Also, the ability of the extract to affect nitric oxide (NO) production in mouse Abelson leukemia RAW264.7 cells and the release of β-hexosaminadase from rat basophilic leukemia (RBL-2H3) cells grown in culture were determined. The ability to inhibit NO production is a measure of the anti-inflammatory effect, while the inhibition of the production of β-hexosaminadase is a measure of the anti-allergic activity. Finally, the cytoactivity of the extract on primary human keratinocytes was evaluated. It inhibited the growth of the Gram-positive bacteria *P. acnes*, *S. aureus* (both standard and isolated strains) and *S. epidermidis*, but not the Gram-negative bacteria *S. typhimurium*, *S. typhi*, *S. sonnei* and *E. coli*. The MIC and MBC values for the extract acting on *P. acnes* were 15.6 and >1000 μg/mL, respectively. The MIC and MBC values for different strains of *S. aureus* and one strain of *S. epidermis* were 7.8–15.6 and 7.8 μg/mL, respectively. Since the MBC values were much higher than the MIC values, the extract was just bacteriostatic and not bacteriocidal" [216]. It inhibited their growth but did not kill them. It is also interesting to note that despite a previous report [217] that ellagitannins isolated from pomegranates had antibacterial activity against both methicillin-resistant and methicillin-sensitive *S. aureus* this more recent study [216] found that these bacteria were not affected by ellagic acid at concentrations up to 2 mg/mL. Still, the inhibitory effects on Gram-positive bacteria may be the reason why pomegranate pericarps have an antidiarrheal effect [216], which was reported by others [218].

In addition, the pomegranate peel extract inhibited the production of nitric oxide [216]. "It took only a concentration of 10.7 μg/mL of the extract to inhibit the production by 50% ($IC_{50}$), which is about the same as the well-known inhibitor of NO synthase, L-nitroarginine. In addition, the extract inhibited the release of β-hexosaminadase; as did ellagic acid. The $IC_{50}$ values were 20.9 and 4.3 μg/mL for the extract and ellagic acid, respectively" [216]. At first, one might think that this means that ellagic acid by itself is more anti-allergenic than pomegranate peels, but this is not supported by analytical data [219]. That is, 20.9 μg/mL of peel polyphenols in the extract contains much less than 4.3 μg/mL ellagic acid. In fact, one important study found that ellagic acid is just one of many phenolic compounds in the peels. That is, there was 638 mg/kg of ellagic acid in dried peels, but a total of 44262 mg/kg total phenolics, including 43979 mg/kg total ellagitannins as well as 447 and 270 mg/kg anthocyanins and hydroxybenzoic acids [219]. Moreover, the phenolic compounds were extracted from dried peels using 80% methanol with 20% water and 0.1% hydrochloric acid at room temperature and pressure [219]. It is quite likely that much more could have been extracted using dry methanol at 100 °C and 10 MPascal pressure, as was seen in a recent analysis of another fruit called noni (*Morinda citrifolia*) [220]. In any case, if the extract used in the study in question [216] had the same relative amount of ellagic acid, it would have been present at only 1.44%. That is, the 20.9 μg/mL extract would have had only 0.30 μg/mL ellagic acid. So, the anti-allergic activity of pomegranate peels must be due to much more than just ellagic acid and there could be some synergism involved. Finally, it should be noted that the use of the extract instead of whole peels was necessary to do the assay of the release of β-hexosaminadase because it is done in an aqueous solution. It is quite possible that the entire peel could have similar or even better health effects *in vivo*.

Pomegranate peels from Yemen were also shown to have antibacterial effects [221]. "The authors focused on bacteria that can cause food-borne illnesses. This included *S. aureus*, *E. coli*, *Listeria monocytogenes*, *S. enteritidis* and *Yersinia enterocolitica*. They tried using three different solvents to prepare extracts from dried peels. They used a Waring blender to thoroughly mix the three portions of dried peels with diethyl ether, 80% methanol or deionized water at room temperature and pressure. After blending, the mixtures were allowed to sit for 1 hr in the dark. Only the 80% methanol (with presumably 20% water) was effective in inhibiting bacterial growth. MICs ranged from 0.25 to 4 μg/mL, with *Y. enterocolitica* being inhibited at 0.25 μg/mL and *S. enteritidis* being 4 μg/mL. Moreover, the 80% methanol extract had an immediate inhibitory effect on the growth of *L. monocytogenes* in fish" [221]. This is important because there was an outbreak of it in the early 1980s [222]. As a final note, 80% methanol most likely will not extract all of the bioactive compounds in pomegranate peels, so the entire peels might be much more effective, thus making them a good food additive for this and many other reasons.

Not only the peels, but also the juice, POMx and a flower extract may have antibacterial and antiviral effects, as described in a recent review [223]. POMx inhibited the growth of *Clostridium* species as well as *S. aureus* [89, 223, 224]. It also stimulated the production of the following probiotic bacteria: *Bifidobacterium breve*, *B. infantis* and *Lactobacillus* species [54, 89, 223]. Pomegranate juice inhibited the growth of *L. monocytogenes* and *S. aureus* [223]. A flower extract inhibited the growth of *Pseudomonas aeruginosa* [223]. The peels inhibited the growth not only the bacteria mentioned in the previous paragraph, but also methicillin-resistant *S. aureus* as well as drug-resistant *Acinetobacter baumannii* and *Helicobacter pylori* [223, 225-228]. Peels also inhibited the growth of the following oral bacteria: *Aggregatibacter actinomycetemcomitans*, *S. aureus*, *S. epidermidis*, *S. mutans*, *S. salivarius*, *S. mitis*, *Porphyromonas gingivalis*, *Prevotella intermedia* and several species of *Proteus* [223, 229-232]. The peels and flowers also inhibited the growth of bacteria that can appear in wounds [223, 234]. Pomegranate fruit also has antiviral effects against the influenza virus, herpes virus, poxviruses, and human immunodeficiency (HIV-1) virus [223, 235-238]. It is also effective against viruses that can cause food-borne illnesses including human noroviruses, hepatitis A virus, rotaviruses, Aichi virus, hepatitis E virus, astroviruses, adenoviruses, parvoviruses, and other human enteroviruses and small round-structured viruses [223, 239]. At doses of 2 and 4 mg/mL, pomegranate juice and polyphenols were also shown to be effective against foodborne viral surrogates (feline calicivirus FCV-F9, murine norovirus MNV-1, and MS2 bacteriophage) [240]. Juice and polyphenols reduced the viral titers (levels of virus) by ≥50% within 20 mins [240].

## ANTIMALARIAL EFFECTS

Pomegranate peels may also be effective in helping to treat malaria [241]. There is an Ayurvedic formulation derived from dried peels called OMARIA (Orissa Malaria Research

Indigenous Attempt) that is used to treat malaria in rural regions of India (especially Orissa) in which the causative agents *Plasmodium falciparum* and *Plasmodium vivax* are found and malaria is endemic [241]. Moreover, strains of *P. falciparum* that are resistant to the antimalarial drug chloroquine have been shown to be sensitive to the toxic action of a tannin-enriched methanolic extract of pomegranate peel powder [241, 242]. The extract inhibited the secretion of matrix metalloprotein-9 (MMP-9) by infected red blood cells [242]. "That is, malaria is driven by the production of cytokines such as tumor necrosis factor, which increases the biosynthesis of MMP-9. The arils from immature pomegranate fruits were removed and the peels that remained were extracted first with petroleum ether (which is mostly hexane) to remove lipids (mostly triglycerides with some plant sterols). The de-lipidized peels were then extracted twice with methanol. Preparative scale liquid chromatography was done to obtain a tannin-enriched fraction. It inhibited the release of MMP-9 that was induced by TNF. Moreover, urolithin metabolites also inhibited MMP-9 secretion" [242]. The use of extracts helped to suggest a mechanism for the anti-malarial effects of the peels, which probably helped to get the paper accepted in a peer-reviewed journal. However, Ayurvedic medicine, like other forms of traditional medicine is based on a holistic approach. So, it is quite unlikely that extracts will ever replace peels in treating malaria in India. This could be important, since there are many more bioactive compounds in the peels than just tannins. Some of them could help increase the bioavailability of the tannins and may act synergistically to help fight malaria.

## USES IN WOUND HEALING

Some of the wound-healing activities of pomegranates have already been mentioned in the section on antibacterial effects. There are at least three other papers that described the wound-healing effects of pomegranates or products made from them [243-245]. In one study, a hydroalcoholic (5% methanol) extract of peels from Tunisian pomegranates was used to prepare an ointment [243]. "It was shown to be able to help heal dermal wounds in guinea pigs. It helped contract the wound and decreased the time needed to generate new epithelial cells. The tensile strength of the wound increased, indicating that it was healing. There was more cellular infiltration, a well-advanced organization of granulation tissue and on-going epithelialization. There was more DNA being synthesized, as well as collagen, the predominant extracellular protein in the granulation tissue of a healing wound. There were no scars after 20 days. The extract also had antibacterial and antifungal activity against *Pseudomonas aeruginosa* ATCC 9027, *S. aureus* ATCC 25923, *E. coli* ATCC 25922, *Klebsiella pneumoniae*, *Salmonella anatum*, *Salmonella typhimurium*, *Streptococcus pneumonia* bacteria and the fungi *Candida albicans*, *C. glabrata*, *Trichopyton rubrum* and *Aspergillus niger*" [243].

In another study, something that the authors called "pomegranate extracts powder" had rapid wound healing effects on surgically created bone defects in the mandible (lower jaw bone) of rabbits [244]. "This was done to mimic periodontal disease. On the first day that the extract was inserted into the wound, there was an acute inflammatory reaction. Many inflammatory cells invaded the surgical area. Inflammatory cells (macrophages eosinophils and neutrophils) invaded the lower border of incision (underlining lamina propria). There was also collagen fiber and congested capillaries. However, by the third day after surgery, the inflamed area was completely separated from underlining normal connective tissue by a basophilic area that closely resembled the appearance of extracellular cartilage. By day 14 the surface was still not fully covered by epithelia, and there were chronic inflammatory cells and eosinophils in the granulation tissue. However, the underlying tissue had active fibroblast and collagen fibers and the tensile strength was significantly greater than the control group. The regeneration of the original tissue was much faster in the skin wound treated with extract ointment" [244].

In the other study [245] ellagic acid and an extract of Chinese pomegranate peels were able to help heal dermal wounds in rats. The powdered extract contained 13% ellagic acid [245]. "The right dorsal surface of each animal was shaved with an electric clipper, and then one linear paravertebral incision 4 cm in length was made with a surgical blade through the full thickness of the skin at 1.5 cm from the midline of the vertebral column" [245]." The wound was closed with three surgical interrupted sutures at equidistant points 1 cm apart" [245]. "A topical formulation of either ellagic acid or the extract was applied to the wound area once daily for 10 consecutive days (from the day of surgery to the 9th day post-surgery). The tensile strength of the healed wound was measured. There was a dose-dependent increase in tensile strength and a contraction of the wound in the animals who received either ellagic acid or the extract. To measure the formation of collagen, the hydroxyproline content was estimated. Hydroxyproline is a characteristic amino acid in collagen. Another group of rats were inflicted with a burn wound. It was immediately treated with 0.5 g of ellagic acid or the extract. The activity of the enzyme myeloperoxidase (MPO) was measured. It is a measure of inflammation. The burn wound contracted and the MPO activity was reduced in the animals who received the extract or ellagic acid. It should be noted that the extract was more efficient than ellagic acid alone. So, there must be other bioactive compounds in the extract that help heal dermal wounds. In conclusion, this study showed that ellagic acid and pomegranate peel extract accelerated cutaneous wound healing in the rat incision, excision and burn wound models" [245].

## USES IN IMPROVING THE ENVIRONMENT

Perhaps the greatest threat to public health and the entire human race is global climate change, provided that we can avoid thermonuclear war. So, the most important health benefits of pomegranates may be those that help the environment. To begin with, the environment is helped by reducing agricultural waste. Peels and other byproducts of pomegranate juice production can be used for a variety of purposes instead of just being placed in landfills where they will decompose and produce greenhouse gases including $CO_2$. In previous sections, the health benefits of solvent extracts of pomegranate peels have been discussed. However, such extracts still leave behind much unused waste since only a small portion of the peels is extracted. Moreover, solvents like methanol, ethanol, hexane and ethyl acetate are greenhouse gases that increase the temperature of the Earth. So, researchers are starting to publish reports about ways to use the peels.

The health benefits of pomegranate peels were reviewed recently [246]. "It focused on their ability to prevent hyperglycemia and to protect the liver. It also described the chemical composition, antioxidant activity and antihyperlipidemic effects of the peels. The mechanisms of their antidiabetic effects were also described. The structures of 26 phenolic compounds found in the peels were shown. This includes valoneic acid dilactone (VAD), the structure of which is shown in Figure 2. As mentioned in previous sections, ellagic acid is also an important bioactive compound in pomegranate peels. The concentration of ellagic acid and the antioxidant activity of 12 different varieties of Iranian pomegranate peels were also reported. The antioxidant activities ranged from 225 to 706 mmmol of Trolox equivalents per 100 g of peels and ellagic acid ranged from 10.0 to 50.0 mg/100 g. The 'Lamsari-e-Behshahr' cultivar had the highest antioxidant activity while the 'Syah-e-Saveh' cultivar had the highest concentration of ellagic acid" [246].

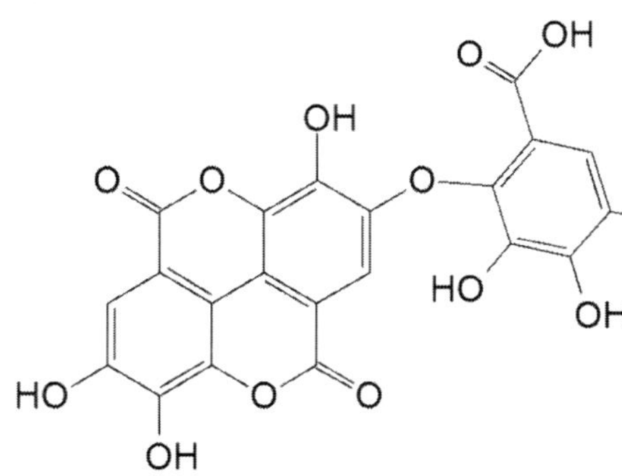


Figure 2. Structure of valoneic acid dilactone (VAD)

VAD was shown to have potent antidiabetic activity in diabetic rats [247]. It prevented damage to the pancreas and inhibited the activities of α-amylase, aldose reductase and protein tyrosine phosphatase 1B with $IC_{50}$ values of 0.284, 0.788, 12.41 μg/mL, respectively [247]. The enzyme α-amylase catalyzes the hydrolysis of alpha bonds in large alpha-linked carbohydrates such as starch and glycogen to produce glucose and maltose. This decreases the concentration of glucose in the blood, which helps to prevent and treat type-2 diabetes. That is, sugars like α-D-glucose can have the hydroxyl group on carbon number 1 in either an alpha (α) of beta (β) position as shown in Figures 3 and 4. By inhibiting α-amylase, VAD and pomegranate peels decrease hyperglycemia and improves dysfunctional glucose metabolism in patients who have type-2 diabetes [248].

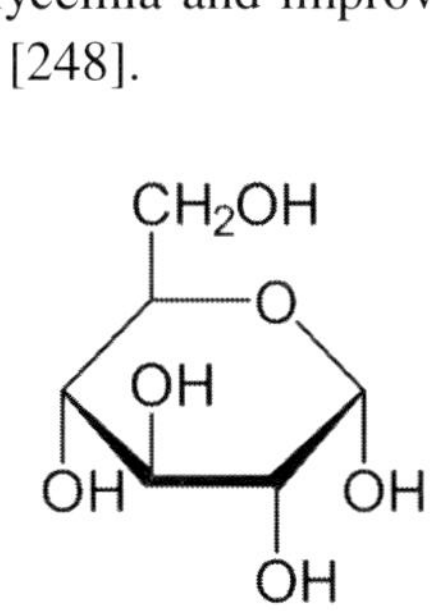


Figure 3. Structure of α-D-glucose.

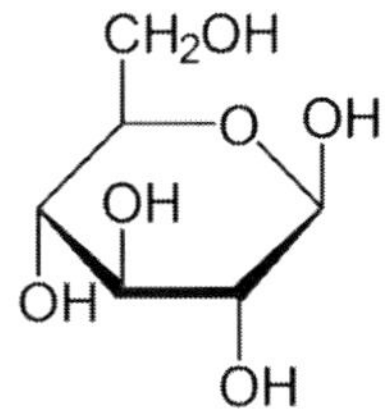


Figure 4. Structure of β-D-glucose.

The enzyme aldose reductase catalyzes the reduction of a variety of aldehydes and carbonyl-containing compounds, including saccharides. It is important in the polyol pathway, which is activated by excess blood glucose and advanced glycation end products that can lead to hypertension, diabetes and cardiovascular diseases. So, by inhibiting aldose reductase, pomegranate peels can help to prevent these diseases. Moreover, aldose reductase plays an important role in diabetic neuropathy, nephropathy, cataracts, retinopathy, and microangiopathy [249]. The enzyme tyrosine phosphatase 1B catalyzes the hydrolysis of phosphate groups in phosphorylated proteins. It is also known as protein tyrosine phosphatase 1B and PTP1B. It inhibits the activity of insulin and is considered to be a potential therapeutic target for type-2 diabetes [247, 250] and is overexpressed in patients who have type-2 diabetes [251]. It is also a potential therapeutic target for breast cancer [252].

As mentioned previously, one of the most important mechanisms by which pomegranate juice prevents many diseases is by the ability of their phenolic compounds to bind $Fe^{2+}$ and $Cu^{2+}$ thus preventing the formation of the highly reactive and dangerous hydroxyl radical [15, 16]. So, an important paper [253] recently described the ability of pomegranate peels to bind not just these, but also other heavy metals that can either undergo the same Fenton reaction and/or are toxic by other mechanisms. This includes cadmium, chromium, manganese, nickel, lead and zinc [253].

Another thing that has been mentioned several times is that extractions of pomegranate peels that are done at room temperature and pressure probably don't solubilize anywhere near all of the bioactive phenolic compounds. So, it is important to note that a recent article described the use of pressurized water to extract them [254]. "The authors found that particle size, temperature, and static time were the most important factors in obtaining a maximum recovery of phenolic compounds. They also found that pressurised water was as effective as using methanol at room temperature. The concentration of total phenolic compounds was 264.3 mg tannic acid equivalents per g of dry weight, compared to 258.2 mg/g for methanol. Hydrolyzable tannins were the predominant polyphenols with a concentration of 262.7 mg/g tannic acid equivalents. The concentration of punicalagin in the peels extracted by pressurised water was 116.6 mg/g dry weight" [254]. So, eco-friendly water should be used instead of the toxic compound and greenhouse gas methanol for preparing pomegranate peel extracts for human consumption. This may also be important in preparing extracts of peels for treating and/or preventing infections by pathogenic organisms such as *Listeria monocytogenes*. A recent study showed that an extract of peels using acetone at 50°C for 48 hrs was effective in damaging the cell membrane of *L. monocytogenes* [255], but acetone is also a greenhouse gas. However, since water can hydrolyze bioactive anthocyanins and most phenolic compounds are more soluble in methanol, extraction with methanol or ethanol at high temperature and pressure is probably better when extracting small amounts of peels for quantitative analysis.

An aqueous extract of pomegranate peels can also be used to prepare dyes for cotton fabrics [256]. "This is much better than using the non-biodegradable chemical dyes that predominate in the clothing and textile industries. Water pollution caused by discharging them is one of the major environmental concerns in the world today. If the primary goal is to help the environment and minimize the cost of production, extraction with water containing 1 g/L of NaCl at 60 °C for one hour can be used. These were the optimum conditions that were determined in the study. Moreover, the dyes stayed in the fabric and were not washed out" [256].

Pomegranate peels can also be used to prepare natural eco-friendly products. One example is using a pomegranate peel extract to prepare nanoparticles as a sustained-release antimicrobial agent [257]. "These nanoparticles are better than conventional drugs that are used topically in the oral cavity but are rapidly cleared by saliva. This causes them to be not as effective as one might want and require multiple administration. So, in dentistry such antimicrobial drugs are often formulated with mucoadhesive polymers" [257]. However, recent studies have shown that nanoparticles have better properties than microparticles because their larger surface area enables them to have more intimate contact with the mucosa [257, 258]. So, a pomegranate peel extract was loaded into mucoadhesive nanoparticles made from polyethyleneimine and dextran sulfate [257]. "The peel extract was added to prevent bad breath (malodor) and cavities (dental caries). They were efficient in that they started releasing the extract within 5 mins and continued releasing it for at least 3 hrs. Moreover, the extract that was loaded into the nanoparticles was as efficient as the free extract in limiting the growth of three oral bacteria, *S. mutan*, *S. sanguinis* and *P. gingivalis*" [257].

Pomegranate peels have also been used to prepare fiber reinforced polymer composites [259]. Such composites can be made at a relatively low cost and are used instead of metals in many industrial applications including aerospace, automotive and boat manufacturing industries [260]. When cars, rockets, cars, trucks and boats are made with light-weight materials, it takes less fossil fuel to propel them, thus producing less carbon dioxide, which is a greenhouse gas. So, composites were made of polyvinyl alcohol, polyacrylamide and pomegranate peels using a casting method [259]. These may have many industrial applications [259].

Composites of pomegranate peels and polystyrene have also been made [261]. "They can be used to prepare materials that can be used as coatings and films for use as dielectric layers, photographic materials and paints. The absorption coefficient, energy gap, refractive index and extinction coefficient all increased with the percentage of pomegranate peels in the composites" [261].

Not only can pomegranate peels be used to prepare polymeric composites, a natural biopolymer, chitosan, can be used to extend the shelf life of pomegranate arils as a postharvest treatment [262]. Chitosan is "a high molecular weight cationic polysaccharide, produced by the deacetylation of chitin, is widely used in postharvest trials because of its excellent film forming and anti-fungal, bio-safe and biochemical properties" [263]. When used in cold storage, chitosan was quite effective in reducing the natural decrease in the concentration of anthocyanins and the color that they provide to the arils [262].

Another byproduct is the solid waste that remains at the bottom of containers used to prepare pomegranate wine [264]. "These are called wine lees and are composed primarily of yeast that binds to polyphenols. They also have relatively high concentrations of polyphenols (about 30 mg GAE/g dry material) and high antioxidant capacities. Analysis of the wine lees found 39 biochemicals, including gallic acid, ellagic acid hexoside, cyanidin 3-glucoside, delphinidin 3-glucoside, cyanidin 3,5-diglucoside, delphinidin 3,5-diglucoside and ellagic acid at concentraitons of 3.27, 2.71, 2.31, 2.10, 1.67, 1.39 and 1.29 mg/g dry material, respectively. So, wine lees might be good food additives" [264].

Such byproducts can also be used to feed animals [265-268]. As the production of pomegranates continues to increase, so does the amount of waste [265]. "The Israeli company Gan Shmuel Food Ltd. (Kibbutz Gan-Shmuel, Israel) has developed a proprietary process to make an extract from the pomegranate parts remaining after producing juice. It was added to the total mixed ration (TMR) of food that is given to lactating cows. It affected the abundance of bacteria that produce methane, another greenhouse gas. It also had a dose-dependent effect on the whole ruminal bacterial community. These changes led to an increase in the digestibility of dry matter, crude protein, and neutral detergent fiber, as well as milk in cows that were fed TMR containing 4% of the extract. So, this could become an important additive to food for cattle" [265].

Another group looked at the effects of adding POMx to the grain that calves ate in their first 70 days of life [266]. "They were fed colostrum during the first 24 hrs, pasteurized milk thereafter until 61 days, and grain *ad libitum* through day 70. The researchers measured their ability to digest nutrients. They also quantified glucose and 3-hydroxybutyrate in their blood serum. Their immunocompetence was also evaluated by quantifying the neutrophil, phagocytic and killing activities as well as antibody response to immunization with ovalbumin. Also, the cytokine production of peripheral blood mononuclear cells was measured. Feeding POMx had no effect on grain intake or gain in body weight the first 30 days. After 30 days, both grain dry matter intake and the rate of body weight gain decreased with increasing addition of POMx. This caused the calves to be 1.8 and 4.3 kg lighter at 70 days of age for doses of five and ten POMx capsules, respectively, when compared with controls. Feeding POMx did not influence the digestibility of dry matter, organic matter, or starch, but it reduced the digestion of crude protein and fat. The concentrations of glucose and 3-hydroxybutyrate in blood plasma were similar among treatments throughout the first 70 days of age. Measures of the health of the calves were not changed by POMx. Similarly, neutrophil phagocytic and killing activities did not differ among treatments" [266].

On the contrary, feeding POMx increased immunocompetence as indicated by increased biosynthesis of interferon-γ and interleukin-4 by peripheral blood mononuclear cells and improved total immunoglobulin G responses to ovalbumin vaccination [266]. "These results suggest that feeding POMx suppresses the intake of grain and digestibility of fat and protein. Nevertheless, polyphenols from POMx enhanced the mitogen-induced cytokine production and response to vaccination, which might benefit immune competence of calves and potentially health. So, more studies are needed to minimize the effect of POMx on the intake and digestibility and to understand better the mechanisms by which polyphenols improve the immune response of calves" [266].

This could be important if POMx or other pomegranate products can be used to minimize or eliminate the need for giving antibiotics to cattle and other animals that are grown in concentrated animal feeding operations (CAFOs) [269]. "All animals living in CAFO are fed antibiotics, with the hope of preventing bacterial infection. Unfortunately, it doesn't work very well. Most of the animals produced by CAFOs contain bacteria. To survive, these bacteria must become resistant to antibiotics. So, the major source of antibiotic resistant bacteria is not people who have been given too many antibiotics, but animals from CAFOs" [269]. Antibiotics tend to give animals (and people) diarrhea. So, CAFO animals, especially cattle, spend their entire lives knee deep in their own fecal material [270]. So, beef is often contaminated with *E. coli* from the gut [270]. The fecal material (especially from hogs) is a major source of environmental pollution. CAFO animals produce much more fecal material and methane than all the humans on our planet [269, 270]. On the other hand, free range cows and other animals not living in CAFOs defecate in the soil, fertilizing it and helping plants grow. Such plants consume carbon dioxide, reducing the production of this greenhouse gas.

Pomegranate seed pulp can also be used in animal feed [267]. "However, they can spoil rapidly. Two ways to prevent this are drying and ensiling (storing in a silo or enclosed pit). It was found that ensiling can increase the nutritive content of the seeds" [267].

Finally, pomegranate peel extracts have been shown to improve fermentation in rumens (like sheep) witht the goal of improving the efficiency of milk and meat production [268]. "These extracts would be much better for the environment than antibiotics, methane inhibitors, and defaunating agents that are currently used. They decreased the production of ammonia and protozoa when added to rumen fluid from sheep in the *in vitro* study. It also improved fermentation as indicated by an increased ratio of propionate to acetate" [268].

There is another potential eco-friendly application for pomegranate peel extracts. They were used to control *Pseudomonas syringae* pv. *tomato*, the causal agent of tomato bacterial speck, which is common in greenhouses and fields all over Italy [271].The antibacterial effect lasted for at least 15 days, permitting the replacement or reduction of the commonly used copper compounds that are not eco-friendly [271].

Pomegranates can also be used to make magnetic hydrogels and nanoparticles for industrial and biomedical applications [272-274]. A magnetic hydrogel was made using pomegranate-like functional magnetic nanospheres (FMNs) to crosslink the polyacrylamide matrix [272]. The FMNs were made from $Fe_3O_4$ nanoparticles embedded into polystyrene [272]. Magnetic hydrogels have potential uses as smart magnetic switches, biomimetic sensors and chemical devices as well as being used in targeted drug release devices [272]. They may be useful in making materials that can mimic or stimulate native tissue functions after organ failure [275]. They are made of a highly hydrated polymeric network that can shaped or casted into different sizes for stem cell engineering and other applications [275]. They can be used as drug delivery systems [276]. Magnetic hydrogels can be used to assemble complex tissue structures [276]. They may also be used to desalinate water by magnetic-field induced heating and forward osmosis [277], as the world continues to look for better ways to produce fresh water, especially as sea levels rises during global climate change.

A pomegranate peel extract can be used to make pomegranate mono-dispersed gold nanoparticles (PAuNPs) [273]. "They added an aqueous solution of the peel extract to 1 mM hydrochloroauric acid ($HAuCl_4$) without using any surfactant or external energy. Physiologically stable, biocompatible PAuNPs were formed within 60 s. Ellagitannins such as punicalagin formed a five-member chelate with the $Au^{3+}$, then reduced it to $Au^0$ as the quinines in the ellagitannins were oxidized. They also served as stabilizing agents, preventing the agglomeration of PAuNPs. Next, the anticancer drug 5-fluorouracil (5-Fu) was coupled to the PAuNPs to make 5Fu@PAuNPs. Then biocompatible casein was added, since is highly expressed in cancer cells. to the prepared PAuNPs. These functionalized 5Fu@PAuNPs were able to release the 5-Fu into MCF-7 breast cancer cells. These nanoparticles are being developed for targeted drug delivery with enhanced therapeutic efficacy and minimal side effects" [273].

A similar study was done in which pomegranate juice was used instead of a peel extract [274]. Gold nanoparticles with interesting catalytic properties were produced [274].

## CONCLUSION

In conclusion, there is much evidence that supports the idea that pomegranate juice and products made from it may have many beneficial health effects. Still, as pointed out in chapter 1, this does not mean that any of the health effects have been proven. So, the courts in the USA and the Federal Trade Commission (FTC) claimed that a manufacturer of pomegranate juice and an extract "deceptively advertised pomegranate products by making unsupported health claims [278]. This is controversial and is discussed in a blog [279].

## REFERENCES

[1] Holland, D. Hatib, K., Bar-Ya'akov. I. *Hort. Rev*. 2009, 35, 127-191.

[2] Seeram, N. P., Schulman, R. N., Heber, D. (eds.). *Pomegranates: Ancient Roots to Modern Medicine*. CRC Press Taylor & Francis Group, Boca Raton, FL, 2006.

[3] Farmahan, H. L. Pomegranate, in *Recent Trends in Horticulture in the Himalayas*, Jindal, K. K., Sharma, R. C. (eds.). p. 139, Indus, New Dehli, India, 2004.

[4] Jayaprakasha, G. K., Negi, P. S., Jena, B. S. Antimicrobial activities of pomegranate," in *Pomegranates: Ancient Roots to Modern Medicine*, Seeram, N. P., Schulman, R. N., Heber, D. eds., p. 168, CRC Press, New York, NY, USA, 2006.

[5] Jurenka, J. *J. Alt. Med. Rev*. 2008, 13, 128-144.

[6] US. FDA, CFR - Code of Federal Regulations Title 21, part 582. Substances Generally Regarded as Safe, 2013. www.accessdata.fda.gov/scripts/cdrh/cfdocs/cfCFR/CFRSearch.cfm?fr=582.20

[7] Cerdá, B. et al. *J. Agric. Food Chem*. 2003, 51, 3493-3501.

[8] Smith, R. E. *Medicinal Chemistry – Fusion of Traditional and Western Medicine*, Bentham Science, Sharjah, U.A.E., 2013.

[9] Seeram, N. P., Lee, R., Heber, D. *Clin. Chim. Acta* 2004, 348, 63-68.

[10] Seeram, N. P. et al. *J. Med. Food* 2008, 11, 390-394.

[11] Mertens-Talcott, S. U et al. *J. Agric. Food Chem*. 2006, 54, 8956-8961.

[12] Kalt, W. et al. *J. Agr. Food Chem*. 2014.

[13] Wayner, D. D. M. et al. *Biochim. Biophys. Acta* 1987, 924, 408-419.

[14] Hallwel, B. *Brit. J. Clin. Pharmacol*. 2012, 75, 637-644.

[15] Kell, D. B. *Arch. Toxicol*. 2010; 84: 825–889.

[16] Sullivan, J. L. *Lancet* 1981, 1, 1293-1294.

[17] Seeram, N. P. et al. *J. Nutr. Biochem*. 2005, 16, 360-367.

[18] Malešević, V. K. et al. *Intl. J. Food Sci. Technol*. 2014, 49, 210-216.

[19] Tezcan, F., Gultekin-Ozguven, M., Diken, T., Ozcelik, B., Erim, B. *Food Chem*. 2009, 873-877.

[20] Lee, C-J., Chen, L-G., Liang, W-L., Wang, C-C. *Food Chem.* 2010, 118, 315-322.
[21] Gil, M. I., Tomás-Barberán, F. A., Hess-Pierce, B., Holcroft, D. M., Kader, A. A. *J. Agric. Food Chem.* 2000, 48, 4581-4589.
[22] Basu, A. et al. *J. Nutr. Metab.* 2013, Article ID 708381, 7 pages.
[23] Çam, M., İçyer, N. C., Erdoğan, F. *LWT-Food Sci. Technol.* 2014, 55, 117-123.
[24] Keşkekoğlu, H. Üren, A. *Meat Sci.* 2014, 96, 1446-1451.
[25] Andjelkovi, M. et al.*Food Chem.* 2006, 98, 23-31.
[26] Benzie, I. F. F., Strain, J. J. *Anal. Biochem.* 1996, 239, 70-76.
[27] Smith, R.E. et al. Noni composition and health benefits, in *Fruit Juices: Types, Nutritional Composition and Health Benefits*, Elder, K.E., ed. Nova Publishers, Hauppauge, NY, 2014.
[28] Richter, B. E. et al. *Anal. Chem.* 1996, 68, 1033-1039.
[29] Aldini, G. et al. *Free Radical Biol. Med.* 2014.
[30] Ahmed, S., Wang, N., Hafeez, B. B., Cheruvu, V. K., Haqqi, T. M. *J. Nutr.* 2005, 135, 2096–2102.
[31] Brosens, I., Pijnenborg, R., Vercruysse, L., Romero, R. *Am. J. Obstet. Gynecol.* 2011, 204, 193–201.
[32] Burton, G. J., Woods, A. W., Jauniaux, E., Kingdom, J. C. *Placenta* 2009, 30, 473–482.
[33] Chen B. et al. *Am. J. Physio.l Endocrinol. Metab.* 2012, 302, E1142–E1152.
[34] Kishore, R. K., Sudhakar, D., Parthasarathy, P.R. *Indian J. Biochem. Biophys.* 2009, 46, 106-11.
[35] Yourtee, D. M., Elkins, D. M., Nalvarte, E. L. Smith, R. E. *Toxicol. Appl. Pharmacol.* 1992, 116, 57-65.
[36] Crocker, I. P., Cooper, S., Ong, S. C., Baker, P. N. *Am. J. Pathol.* 2003, 162, 637-643.
[37] Leung, D. N. et al. *Am. J. Obstet. Gynecol.* 2001, 184, 1249–1250.
[38] Smith, S. C., Baker, P. N., Symonds, E. M. *Am. J. Obstet. Gynecol.* 1997, 177, 1395-1401.
[39] Ahmed, S., Wang, N., Hafeez, B. B., Cheruvu, V. K., Haqqi, T. M. *J. Nutr.* 2005, 135, 2096–2102.
[40] Nagase, H., Woessner, Jr. J. F. Matrix metalloproteinases. *J. Biol. Chem.* 1999, 274, 21491-21494.
[41] Ahmed, S., Wang, N., Lalonde, M., Haqqi, T. M. *J. Pharmacol. Exp. Ther.* 2004, 308, 767–773.
[42] Al-Muammar, M. N. *Nutr.* 2012, 28, 595-604.
[43] Vroegrijk, I. O. et al. *Food Chem. Toxicol.* 2011, 49, 1426–30.
[44] Hontecillas, R. et al. *J. Am. Coll. Nutr.* 2009, 28, 184–95.
[45] Lei, F., et al. *Int. J. Obes. (Lond).* 2007, 31, 1023–9.
[46] Aviram, M. et al. *J. Agr. Food Chem.* 2008, 56, 1148-1157.
[47] Li, Y. Et al. *Diabetes Obes. Metab.* 2008, 10, 10–7.
[48] Liu, J. *J Ethnopharmacol.* 2005, 100, 92–4.
[49] De Melo, C. L. et al. *Chem. Biol. Interact.* 2010, 185, 59–65.
[50] Jang, A. et al. *Chem. Biol. Interact.* 2008, 174, 109–17.
[51] Xu, K. Z. et al. *J. Ethnopharmacol.* 2009, 123, 280-287.
[52] Cerda, B. et al. *Eur. J. Nutr.* 2003, 42, 18–28.
[53] Esmaillzadeh, A. et al. *Int. J. Vitam. Nutr. Res.* 2006, 76, 147–51.

[54] Bialonska, D. et al. *J. Agric. Food Chem.* 2009, 57, 8344–9.
[55] Larrosa, M. et al. *J. Nutr. Biochem.* 2009, 21, 717–25.
[56] Kim, H.-S. et al. *Exp. Biol. Med.* 2014.
DOI: 10.1177/1535370213514511.
[57] Kim, H.-S. et al. *Nutr. Res.* 2013, 33, 144-153.
[58] Zou, X. et al. *Antioxid. Redox Signaling* 2014.
[59] Abozid, M. M., Farid, H. E. A. *J. Biol. Chem. Environ. Sci.* 2013, 8, 83-104.
[60] Adeghate, E. *Curr. Med. Chem.* 2008, 15, 1851-62.
[61] Al-Suhaimi, E., Shehzad, A. *Eur. J. Med. Res.* 2013, 18(12), 1-13.
[62] Koda, M. et al. *BMC Cancer*, 2010, 10, 320.
[63] Romacho, T., Sánchez-Ferrer, C. F., Peiro, C. *Mediat. Inflamm.* 2013, Article ID 946427, 15 pages.
[64] Nogeuira, A. V. et al. *Clin. Oral Invest.* 2014, 18, 171-178.
[65] Colombo, E., Sangiovani, E. Dell'Agli, M. *EMB Compl. Alt. Med.* 2013, Article ID 247145, 11 pages.
[66] Saniee, P. et al. *Helicobacter* 2009, 14, 393-394.
[67] Hajimahmoodi, M. et al. *Nat. Prod. Res.* 2011, 25, 1059-1066.
[68] Voravuthikunchai, S. P., Mitchell, H. *J. Health Sci.* 2008, 54, 81-88.
[69] Ajaikumar, K. B., Asheef, M., Babu, B.H., Padikkala, J. *J. Ethnopharmacol.* 2005, 96, 171-176.
[70] Alam, S. M. et al. *Toxicol. Mech. Meth.* 2010, 20, 572-578.
[71] DeKosky, S. T. et al. *J. Amer. Med. Assoc.* 2008, 300, 2253-2262.
[72] Gharzouli, K., Khennouf, S., Amira, S., Gharzouli, A. *Phytother. Res.* 1999, 13, 42-45.
[73] Beserra, A. M. S. E. S. et al. *J. Agr. Food Chem.* 2011, 59, 6957-6965.
[74] Wallace, J. L. *Physiol. Rev.* 2008, 88, 1547-1565.
[75] Lai, S. et al. *Zhongguo Zhongyao Zazhi* 2009, 34, 1290-1294.
[76] Gautam, R., Sharma, S.C. *Intl. J. Pharm. Sci.* 2012, 4, 451–461.
[77] Murakami, S. et al. *Planta Med.* 1991, 57, 305-308.
[78] Iino, T. et al. *Life Sci.* 2002, 70, 1139-1150.
[79] Adams, L. S. et al. *J. Agr. Food Chem.* 2006, 54, 980-985.
[80] Leslie, N. R., Brunton, V. G. *Sci.* 2013, 341, 355-356.
[81] Romier-Crouzet, B. et al. *Food Chem. Toxicol.* 2009, 47, 1221-1230.
[82] SA Biosciences, Pathways online,
www.sabiosciences.com/pathwaysonline/
[83] Artursson, P., Palm, K., Luthman, K. *Adv. Drug Delivery Rev.* 2001, 46, 27-43.
[84] Hollebeeck, S. et al. *Food Function* 2012, 3, 875-885.
[85] Scalbert, A., Williamson, G. *J. Nutr.* 2000, 130, 2073S–2085S.
[86] Parkar, S. G.; Stevenson, D. E.; Skinner, M. A. *Int. J. Food Microbiol.* 2008, 124, 295–298.
[87] Seeram, N.P. et al. *J. Nutr.* 2006, 136, 2481–2485.
[88] Chung, K. T.; Lu, Z.; Chou, M. W. *Food Chem. Toxicol.* 1998, 36, 1053–1060.
[89] Bialonska, D. et al. *Intl. J. Food Microbiol.* 2010, 140, 175-182.
[90] Rastall, R. A. et al. *FEMS Microbiol. Ecol.* 2005, 52, 145–152.
[91] Sergent, T. et al. *Chem. Biol. Interact.* 2010, 188, 659–667.
[92] González-Sarrías, A., Espín, J. C., Tomás-Barberán, F. A.,
García-Conesa, M. T. *Molec. Nutr. Food Res.* 2009, 53, 686–698.

[93] González-Sarrías, A., Larrosa, M., Tomás-Barberán, F. A., Dolara, P., Espín, J. C. *Brit. J. Nutr.* 2010, 104, 503–512.
[94] Gimenez-Bastida, J. A. et al. *J. Agr. Food Chem.* 2012, 60, 8866–8876.
[95] Ogawa, Y. et al. *Life Sci.* 2002, 71, 827–839.
[96] Singh, K., Jaggi, A. S., Singh, N. *Phytother. Res.* 2009, 23, 1565–1574.
[97] Umesalma, S., Sudhandiran, G. *Basic Clin. Pharmacol. Toxicol.* 2010, 107, 650–655.
[98] Rosillo, M. A., Sanchez-Hidalgo, M., Cárdeno, A., de la Lastra, C. A. *Biochem. Pharmacol.* 2011, 82, 737–745.
[99] Rosillo, M. A. et al. *Pharmacolog. Res.* 2012, 66, 235–242.
[100] Lian, J., Ding, W., Sun, J. X., Liu, X. M. *Pharmaceut. Care Res.* 2009, 9, 107–110.
[101] Neyrinck, A.M. et al. *Brit. J. Nutr.* 2012, 7, 1–8.
[102] Khokhlova, E. V. et al. *Microbiol. Immunol.* 2012, 56, 27–39.
[103] Larrosa, M. et al. *J. Nutr. Biochem.* 2010, 21, 717–725.
[104] Boussetta, T. et al. *PLoS ONE*, 2009, 4, Article ID e6458.
[105] Bassaganya-Riera, J. et al. *Brit. J. Nutr.* 2011, 106, 878–886.
[106] Katayama, K. et al., *Gastroenterol.* 2003, 124, 1315–1324.
[107] Coursodon-Boyiddle, C. F. et al., *Am. J. Physiol.* 2012, 303, G744–G751.
[108] Hasan, R. et al., *Latin Am. J. Pharm.* 2009, 28, 783–788.
[109] Qnais, E. Y., Elokda, A. S., Ghalyun, Y. Y. A., Abdulla, F. A. *Pharmaceut. Biol.* 2007, 45, 715–720.
[110] Katz, S.R., Newman, R.A., Lansky, E.P. *J. Med. Food* 2007, 10, 213-217.
[111] Banihani, S., Swedan, S., Alguraan, Z. *Nutr. Res.* 2013, 33, 341-348.
[112] Huang, T. H. W. et al. *J. Cardiovasc. Pharmacol.* 2005, 46, 856-862. [113] Wang, J., Rong, X., Um, I.S., Yamahara, J., Li, Y. *EBC Alternat. Med.* 2012, Article ID 350125.
[114] Huang, T. H. et al. *Toxicol. Appl. Pharmacol.* 2005, 207, 160–9.
[115] Jafri, M. A., Aslam, M., Javed, K., Singh, S. *J. Ethnopharmacol.* 2000, 70, 309–14.
[116] Huang, T. H. et al. *Br. J. Pharmacol.* 2005, 145, 767–74.
[117] Pirbalouti, A. G., Azizi, S., Koohpayeh, A., Hamedi, B. *Acta Pol. Pharm.* 2010, 67, 511–6.
[118] Cambay, Z., Baydas, G., Tuzcu, M., Bal, R. *Acta Physiol. Hung.* 2011, 98, 409–20.
[119] Castellano, J.M. et al. *Diabetes*, 2013, 62, 1791-1799.
[120] Li, Y. et al. *Toxicol. Appl. Pharmacol.* 2014, 277, 155-163.
[121] De Melo, C.L. et al. *Chem.-Biol. Interact.* 2010, 185, 59-65.
[122] Das, A. K. et al. *Phytother. Res.* 2001, 15, 628–629.
[123] McFarlin, B. K., Strohacker, K. A., Kueht, M. L. *Br. J. Nutr.* 2009, 102, 54–9.
[124] Khalil, E. A. M. *Egyptian J. Hosp. Med.* 2004, 16, 92–99.
[125] Parmar, H.S., Kar, A. *J. Med. Food* 2008, 11, 376–81.
[126] Rock, W. et al. *J. Agric. Food Chem.* 2008, 56, 8704–13.
[127] Esmaillzadeh, A. et al. *Med. Food* 2004, 7, 305–708.
[128] Koren-Gluzer, M., Aviram, M., Meilin, E., Hayek, T. *Atherosclerosis* 2011, 219, 510–8.
[129] Makino-Wakagi Y, et al. *Biochem. Biophys. Res. Commun.* 2012, 417, 880–5.
[130] Steppan, C. M. *Nature* 2001, 409, 307–312.
[131] Young, B. S. et al. *J. Clin. Endocrinol. Metab.* 2004, 89, 150–156
[132] Ahmad, H. I., Al-Tai, S. K. *J. Adv. Biomed. Pathobiol. Res.* 2013, 3, 6-17.

[133] Satomi, H. et al. *Biol. Pharm. Bull.* 1993, 16, 787-790.
[134] Supuran, C. T. *Bioorg. Med. Chem. Letters* 2010, 20, 3467-3474.
[135] Corey, E. J., Czako, B., Kürti, L. *Molecules and Medicine*, John Wiley & Sons, New York, 2007.
[136] Hakan, P. et al. *J. Pineal Res.* 2003, 35, 85-90.
[137] Ghosh, Ghosh, D., Scheepens, A. *Mol. Nutr. Food Res.* 2009, 53, 322 – 331.
[138] Stowe, C. B. *Complement Ther. Clin. Pract.* 2011, 17, 113-115.
[139] Aviram, M., Dornfield, L. *Atherosclerosis* 2001, 158, 195-198.
[140] van Greevenbroek, W., Schalkwijk, C. G., Stehouwer, C. D. A. Neth. J. Med. 2013, 71, 174-187.
[141] De Nigris, F. et al. *Proc. Natl. Acad. Sci.* 2005, 102, 4896-4901.
[142] Nipcon, E., Marieb, H. *Human Anatomy and Physiology*, Pearson College Division, London, 2012.
[143] Buga, G. M., Gold, M. E., Ignarro, L. J. (1991) *Hypertension* 17, 187–193.
[144] De Nigris, F. et al. *Trends Mol. Med.* 9, 351–359.
[145] Gimbrone, M. A., Jr. (1999) *Am. J. Pathol.* 155, 1–5.
[146] Rozenberg, O., Howell, A., Aviram, *Atherosclerosis* 2006, 188, 68-76.
[147] Mohan M, Waghulde H, Kasture S. *Phytother. Res.* 2010, 24, S196-203.
[148] Aviram M, et al. *Drug.Exp. Clin. Res.* 2001, 28, 49-62.
[149] Aviram M, et al. *Clin. Nutr.* 2004, 23, 423-433.
[150] Ito, H. et al. *Food Chem.* 2014, 152, 323-330.
[151] Nagai, R. et al. Anti-Aging Medicine, 2010, 7, 112–119.
[152] Mayo Clinic. Arteriosclerosis/atherosclerosis. www.mayoclinic.org/diseases-conditions /arteriosclerosis-atherosclerosis/basics/definition/con-20026972
[153] Schubert, S. Y., Neeman, I., Resnick, N. *FASEB J.* 2002, 16, 1931–3.
[154] Aggarwal, B. B., Shishodia, S. *Ann. N. Y. Acad. Sci.* 2004, 1030. 434–41.
[155] Hamoud, S. et al. *Atherosclerosis* 2014, 232, 204-210.
[156] Huang, T. H.-W. eta l. *Brit. J. Pharmacol.* 2005, 145, 767-774.
[157] Azadzol, K. M. et al. *J. Urol.* 2005, 174, 386-393.
[158] Wang, T. et al. Int. J. Impot. Res. 2004, 16, 403-411.
[159] Lansky, E. P., Newman, R.A. *J. Ethnopharmacol.* 2007, 109, 177-206.
[160] Adhami, V.M., Khan, N., Mukhtar, H. *Nutr. Cancer* 2009, 61, 811-815.
[161] Bell, C., Hawthorne, S. *J. Pharm. Pharmacog.* 2008, 60, 139-144.
[162] Sreekumar, S. et al. *BioMed. Res. Intl.* 2014, Article ID 686921, 12 pages.
[163] Bekir, J., Mars, M., Souchard, J. P., Bouajila, J. *Food Chem. Toxicol.* 2013, 55, 470-475.
[164] Kim, N. D. et al. *Breast Cancer Res. Treatment* 2002, 71, 203-217.
[165] Shirode, A.B. *Mol. Carcinogen.* 2014, 53, 458-470.
[166] Mehta, R., Lansky, E. P., 2004, *Eur. J. Cancer Prevent.* 13, 345-348.
[167] Hora, J. J., Maydew, E. R., Lansky, E. P., Dwivedi, C., 2003, *J. Med. Food* 6, 157-161.
[168] Afaq, F. et al. *Intl. J. Cancer* 2005, 113, 423-433.
[169] Afaq, F. et al. *Exp. Dermatol.* 2009, 18, 553-561.
[170] Pacheco-Palencia, L. A. et al. *J. Agr. Food Chem.* 2008, 56, 8434-8441.
[171] Syed, D. N. et al. *Photochem. Photobiol.* 2006, 82, 398-405.
[172] Kohno, H. et al. *Cancer Sci.* 2004, 95, 481-486.
[173] Adams, L. S. et al. *J. Agric. Food Chem.* 2006, 54, 980-985.

[174] Banerjee, N. et al. *Carcinogenesis* 2013, 34, 2814-2822.
[175] Thakur, V. S. et al. *AAPS J.* 2014, 16, 151-163.
[176] Seeram, N. P. et al. *J. Nutr. Biochem.* 2005, 16, 360-367.
[177] Karaaslan, M., Vardin, H., Suzan Varliköz, S., Yılmaz, F. M. *Intl. J. Food Sci. Technol.* 2014, 49, 82-90.
[178] Bhatia, D. et al. *EBC Alt. Med.* 2013, Article ID 371813, 15 pages.
[179] Bishayee, A. et al. *Carcinogenesis* 2011, 32, 888-896.
[180] Tran, H. N. A. et al. *Endocrine Res.* 2010, 35, 1-16.
[181] Lee, S.-T. et al. *BMC Compl. Alt. Med.* 2013, 13:364, 11 pp.
[182] Bell, C., Hawthorne, S. *J. Pharm. Pharmacog.* 2008, 60, 139-144.
[183] Schubert, S. Y., Lansky, E. P., Neeman, I. *J. Ethnopharmacol.* 1999, 66, 11–17.
[184] Heber, D. et al. *J. Agric. Food Chem.* 2007; 55, 10050-10054.
[185] Kroeger, N., Belldegrun, A. S., Pantuck, A. J. *EBC Alt. Med.* 2013; Article ID 701434, 9 pages.
[186] Zhang, Q., Radisavljevic, Z. M., Siroky, M. B., Azadzoi, K. M. *Intl. J. Androl.* 2010, 33, 1-11.
[187] Azadzoi, K. M., Schulman, R. N., Aviram, M., Siroky, M. B. *J. Urol.* 2005, 174, 386-393.
[188] Albrecht, M. et al. *J. Med. Food* 2004, 7, 274-283.
[189] Malik, A. et al. *Proc. Natl. Acad. Sci.* 2005, 102, 14813-14818.
[190] Chiao, J. W. et al. *Carcinogenesis* 2004, 25, 1403–1408.
[191] Sartippour, M. R. et al. *Intl. J. Oncol.* 2008, 32, 475-480.
[192] Hong, M. Y., Seeram, N. P., Heber, D. *J. Nutr. Biochem.* 2008, 19, 848-855.
[193] Ming, D.-S. et al. *J. Steroid Biochem. Mol. Biol.* 2014, 143, 19-28.
[194] Staridi, F., Karapanagiotou, E. M., Syrigos, K. N. *Cancer Treatment Rev.* 2010, 36, 122–130.
[195] Seeram, N.P. et al. *J. Nutr. Biochem.* 2005, 16, 360-367.
[196] Lansky, E. P. et al. *Investigational New Drugs* 2005, 23, 11-20.
[197] Seeram, N.P. et al. *J. Agric. Food Chem.* 2007, 55, 7732-7737.
[198] Pantuck, A. J. et al. *Clin. Cancer Res.* 2006, 12, 4018-4026.
[199] Ropacki, S. A., Patel, S. M., Hartman, R. E. *EBC Alt. Med.* 2013, Article ID 932401, 8 pp.
[200] Hartman, R.E. et al. *Neurobiol. Dis.* 2006, 24, 506-515.
[201] Rojanathammanee, L., Puig, K. L., Colin K., Combs, C. K. *J. Nutr.* 2013, 143, 597-605.
[202] Jahromy, M. H., Khakpour, S., Khorgami, Z. *Chin. Med.* 2014, 5, 1-6.
[203] Dulcich, M. S., Hartman, R. E. *EBC Alt. Med.* 2013, Article ID 940830, 8 pp.
[204] Tapias, V., Cannon, J. R., Greenamyre, J. T. *Neurobiol. Aging* 2013. [205]Caesar & Loretz GmbH, Miglyol® 812 Material Safety Data Sheet, 2013.
[206] Caparros-Lefebvre, D., Elbaz, A. *Lancet* 1999, 354, 281-285.
[207] Escobar-Khondiker, M. et al. *J. Neurosci.* 2007, 27, 7827–7837.
[208] Lannuzel, A. et al. *Brain* 2007, 130, 816-827.
[209] Höllerhage, G. U. et al. *Exp. Neurol.* 2009, 220, 133–142.
[210] Potts, L. F. et al. 2012, *Neurotoxicol.* 33, 53-58.
[211] Champy, P. et al. *J. Neurochem.* 2004, 88, 63–69.
[212] Liaw, C. C. et al. *Angew. Chem. Int. Ed.* 2011, 50, 7885-7891.

[213] Février, A. et al. *Planta Medica* 1999, 65, 47-49
[214] Gu Z.-M. et al. *Phytochem. Anal.* 1999, 10, 32–38.
[215] Smith, R.E., Tran, K., Richards, K. M. Bioactive annonaceous acetogenins, in *Studies in Natural Products Chemistry*, Elsevier, New York, 2014.
[216] Panichayupakaranant, P., Tewtrakul, S., Yuenyongsawad, S. *Food Chem.* 2010, 123, 400-403.
[217] Machado, T. B.et al. *Int. J. Antimicrob. Ag.* 2003, 21, 279–284.
[218] Das, A. K. et al. *J. Ethnopharmacol.* 1999, 68, 205-208.
[219] Fischer, U. A., Carle, R., Kammerer, D. R. *Food Chem.* 2011, 127, 807-821.
[220] Smith, R.E., Tran, K., Richards, K. M. Noni composition and health benefits, in *Studies in Natural Products Chemistry*, Elsevier, New York, 2014.
[221] Al-Zoreky, N. S. *Int. J. Food Microbiol.* 2009, 244-248.
[222] Cornu, M. et al. *J. Food Microbiol.* 2006, 106, 159-168.
[223] Howell, A. B., D'Souza, D. H. *EBC Alternative Med.* 2013, Article ID 606212, 11 pages.
[224] Tayel, A. et al. *Foodborne Path. Dis.* 2012, 9, 755-761.
[225] Gould, S. W. J. et al. *BMC Compl. Alt. Med.* 2009, 9, article 23, 2009.
[226] Phatthalung,, P. N., Chusri, S., Voravuthikunchai, S. P. *BMC Compl. Alt. Med.* 2012, 12, article 56, 8 pages.
[227] Hajimahmoodi, M. et al. *Nat. Prod. Res.* 2011, 25, 1059-1066.
[228] Abdollahzadeh, S. et al. *J. Dentistry* 2011, 8, 1-6.
[229] Vasconcelos, L. et al. *Braz. Dental J.* 2006, 17, 223-227.
[230] Rosas-Piñon, Y. et al. *J. Ethnopharmacol.* 2012, 141, 860-865.
[231] Bhadbhade, S. J. et al. *Quintessence Int.* 2011, 42, 29-36.
[232] Menezes, S. M. S., Cordeiro, L. N., Viana, G. S. B. *J. Herbal Pharmacol.* 2006, 6, 79-92.
[233] Hayouni, E. A. et al. *Phytomed.* 2011, 18, 976-984.
[234] Pirbalouti, A. G. et al. *Acta Poloniae Pharmaceutica*, 2010, 67, 511–516.
[235] Pirbalouti, A. G. et al. *Acta Poloniae Pharmaceutica*, 2010, 67, 107-110.
[236] Haidari, M., Ali, M., Casscells, S. W., Madjid, M. *Phytomed.* 2009, 16, 1127-1136.
[237] Neurath, R. et al. *Punica granatum* (pomegranate) juice provides an HIV-1 entry inhibitor and candidate topical microbicide," in *Natural Products and Molecular Therapy*, G. J. Kotwal and D. K. Lahiri, Eds., vol. 1056, pp. 311–327, New York Academy of Sciences, 2005.
[238] Kotwal, G. J. *Vaccine* 2008, 26, 3055-3058.
[239] D'Souza, D. H. et al. *Int. J. Food Microbiol.* 2006, 108, 84-91.
[240] Su, X., Sangster, X., D'Souza, D. H. *Foodborn. Path. Dis.* 2011, 8, 1177-1183.
[241] Ismail, T., Sestili, P., Akhtar, S. *J. Ethnopharmacol.* 2012, 143, 397-405.
[242] Dell'Agli, M. et al. *J. Ethnopharmacol.* 2009, 125, 279-285.
[243] Hayouni, E.A. *Phytomed.* 2011, 18, 976-984.
[244] Qassab, Z. A. *J. Biol.* 2014, 4, 102-114.
[245] Mo, J. et al. *J. Nat. Med.* 2014, 1-10.
[246] Middha, S. K., Usha, T., Pande, V. *EBC Alt. Med.* 2013, Article ID 656172, 10 pages.
[247] Jain, V. et al. *EBC Alt. Med.* 2012, Article ID 147202, 11 pages.
[248] Punitha, R., Manoharan, S. *Ehtnopharmacol.* 2006, 105, 39-46.
[249] Scotti, L. et al. *Braz. J. Pharmacog.* 2011, 21, 170-180.

[250] Combs, A. P. *J. Med. Chem.* 2010, 53, 2333–44.
[251] Goldstein, B. J. *J. Biol. Chem.* 2000, 275, 4283-4289.
[252] Lessard L, Stuible M, Tremblay ML. *Biochim. Biophys. Acta* 2010, 1804, 613–9.
[253] Ozcan, M. M., Dursun, N. Sağlam, C. *Int. J. Food Prop.* 2011, 14, 550-556.
[254] Cama, M., Hisil, Y. *Food Chem.* 2010, 123, 878-885.
[255] Li, G. et al. *Foodborn. Path. Dis.* 2014, 11, DOI: 10.1089/fpd.2013.1675.
[256] Adeel, S. et al. *Asian J. Chem.* 2009, 21, 3493-3499.
[257] Tiyaboonchai, W., Rodleang, I., Ounaroon, A. *Pharm. Dev. Technol.* 2014, 1–7.
[258] Chaturvedi, M., Kumar, M., Pathak, K. J. *Adv. Pharm. Technol. Res.* 2011, 2, 215–222.
[259] Hashim, A., Husaien, M., Ghazi, J. H., Hakim, H. *Universal J. Phys. Application* 2013, 1, 242-244.
[260] Devendra1, K., Rangaswamy, T. *J. Comp. Eng. Res.* 2012, 2, 1708-1714.
[261] Jasim, F. A. et al. *Adv. Physics Theories Appl.* 2013. ISSN 2224-719X.
[262] Varasteh, F., Arzani, K., Barzegar, M., Zamani, Z. *Food Chem.* 2012, 130, 267-272.
[263] Lin, L. et al. *J. Food Agric.* 2008, 88, 877-884.
[264] Mena, P. et al. *Food Chem.* 2014, 145, 327-334.
[265] Jami, E. et al. *J. Dairy Sci.* 2012, 95, 1–10.
[266] Oliveira, R. A. et al. *J. Dairy Sci.* 2010, 93, 4280-4291.
[267] Taher-Maddah, M., et al. *Open Vet. J.* 2012, 2, 40-45.
[268] Abarghuei, M. J., Rouzbehani, Y., Salem, A.-F. *Turkish J. Vet. Animal Sci.* 2014, 38, 1-8.
[269] Kingsolver B. *Animal, Vegetable, Miracle.* Harper Perennial, New York, 2007.
[270] Schlosser E. *Fast Food Nation.* Houghton-Mifflin, New York, 2001.
[271] Quattrucci, A. et al. *Crop Protection* 2013, 46, 18-22.
[272] Xia, M. et al. *Aust. J. Chem.* 2014, 67, 112-120.
[273] Ganeshkumar, M. et al. *Colloids Surf. B.* 2013, 106, 208-216.
[274] Dash, S. S., Braja Gopal Bag, B. G. *Appl. Nanosci.* 2014, 4, 55-59.
[275] Gaharwar, A. K. Peppas, N. A., Khademhosseini, A. *Biotechnol. Bioeng.* 2014, 111, 441-452.
[276] Lee, S. C., Kwon, I. K., park, K. *Adv. Drug Del. Rev.* 2013, 65, 17-20.
[277] Razmjou, A. et al. *Env. Sci Technol.* 2013, 47, 6297-6305.
[278] US Federal Trade Commission. www.ftc.gov/news-events/press-releases/2013/01/ ftc-commissioners-uphold-trial-judge-decision-pom-wonderful-llc
[279] US FDA. www.ftc.gov/news-events/press-releases/2013/01/ftc-commissioners-uphold-trial-judge-decision-pom-wonderful-llc

# APPENDIX

## CHEMISTRY AND BIOCHEMISTRY

As mentioned in chapter 4, there are triglycerides, fatty acyls, terpenes, alkaloids, phenolic compounds, tannins, organic acids, flavonols, anthocyanins and anthocyanidins and procyanidins in pomegranates. Triglycerides are the main ingredient in oils obtained from pomegranates and other fruits and vegetables. They are esters formed by the reaction between glycerol and three fatty acids, as shown in Figure 8 in Chapter 1. All fatty acids have a slightly acidic carboxylic acid (COOH) group, as shown in Figure 1.

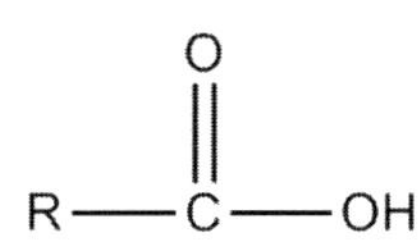

Figure 1. Generalized structure of a fatty acid, where R is a hydrocarbon chain.

When a fatty acid reacts with a hydroxyl group on glycerol or glycerol 3-phosphate, it loses its own –OH while the hydroxyl on the glycerol loses its H to form a molecule of water and a monoacylglycerol or monoacylglycerol phosphate, also known as a lysophospholipid. The fatty acid is converted to a fatty acyl, the generalized structure of which is shown in Figure 2.

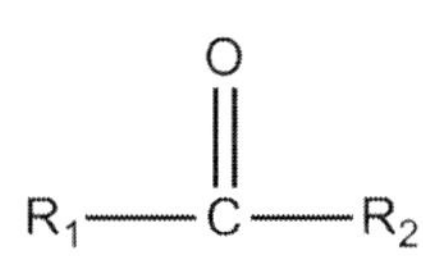

Figure 2. Generalized structure of a fatty acyl, where $R_1$ is a hydrocarbon chain and $R_2$ can be any of a number of things, including HC-O- or $H_2C$-O- of a glycerol backbone or coenzyme S, for fatty acyl CoA.

The reaction of a second fatty acid with a monoacylglycerol or lysophospholipid will form another water molecule and a diacylglycerol or phospholipid. The reaction between a third fatty acid and a diacylglycerol will form a third water molecule and a triglyceride. Actually, the fatty acids themselves don't react directly with glycerol. Instead, they are activated by being converted to an ester of coenzyme A, abbreviated as CoA and CoASH, because it has a sulfhydryl –SH group. When it reacts with a fatty acid, it makes a thioester, often abbreviated as fatty acylCoA. When the acid is acetic acid, it is acetylCoA, which is a major hub in the metabolic network of the cell as it feed into the tricarboxylic acid cycle to produce energy [1]. So, the reactions between acylCoA and either glycerol, monacyl- or diacylglycerol are catalyzed by a class of enzymes called acyl transferases [2]. This occurs when a cell wants to store energy in the form of triglycerides. Triglycerides form large, oily droplets that are stored in white adipose tissue in humans and animals [1]. When nutritional sources of energy are scarce, the triglycerides in the lipid droplet are hydrolyzed, generating fatty acids that are exported into the blood and circulate to other tissues that need energy. This reaction is catalyzed by a class of enzymes called lipases. Fatty acids can be further broken down to produce energy in a process called β-oxidation [1].

So, the blood contains fatty acids and triglycerides [1]. Having an excess (>150 mg/dL) of triglycerides in the blood is a risk factor for heart disease and strokes. That is, hypertriglyceridemia is diagnosed by measuring the fasting level of triglycerides in the blood. Mild and moderate hypertriglyceridemia (150–999 mg/dL) increase the risk of cardiovascular diseases. Severe and very severe hypertriglyceridemia (triglycerides > 1000 mg/dL) are also considered to be a risk for pancreatitis.

Blood also contains red and white blood cells that have cell membranes made of phospholipids, cholesterol, proteins, glycoproteins and lipoproteins [1]. All of the human cells in our bodies also contain these, too [1]. There are also internal organelles (except in red blood cells) that have their own membranes that contain these biopolymers [1].

Adipose cells store triglycerides. They can be either white or brown [1, 3]. "Brown adipocytes have more mitochondria than white adipocytes. Enzymes in brown adipocytes will catalyze the hydrolysis of the triglycerides to make energy for themselves. Adipose tissue not only stores energy, but also secretes hormones as part of the endocrine system. One of them is leptin. It goes to the brain and it causes one to feel full and stop eating. It plays a crucial role in regulating energy intake and expenditure. Leptin-mediated signals from the brain regulate lipid metabolism in adipocytes. During starvation, the concentration of leptin falls, signaling a decline in energy stores in the brain. This causes changes in the expression of genes in the hypothalamus that affects thyroid hormone levels [1, 4]. Unfortunately for modern humans who have ready access to food, the opposite does not occur. When leptin concentrations increase (after eating), we develop a resistance to leptin and gain weight [4].

Brown fat also contains subcellular particles called peroxisomes [1]. Their main function is to break down fatty acids by β-oxidation. "Peroxisomes have a receptor called a peroxisome proliferator-activated receptor, or PPAR. When PPARs are activated, they induce the transcription of several insulin-responsive genes. These genes help control glucose and lipid metabolism. The PPAR family is composed of three proteins: PPAR-α, PPAR-δ (also known as PPAR-β) and PPAR-γ. They are nuclear receptors, which interact with other proteins. The PPARs form heterodimers with the retionid X receptor, or RXR. Most PPAR/RXR heterodimers are constitutively bound to PPAR response elements (PPREs) that are in the promoter regions of target genes. When a ligand binds, it causes a conformational change that causes the degradation of co-repressor complexes, the recruitment of coactivator complexes, and the subsequent induction of target-gene expression. PPAR-α was the first isoform that was discovered. It is the target of the prescription hypolipidemic fibrate drugs and other compounds that induce peroxisome proliferation. One of the most important classes of endogenous ligands of PPARs is unsaturated fatty acids (not triglycerides)" [1]. So, it is important to properly distinguish between fatty acids and fatty acyls that are a part of triglycerides.

It is also important to be able to distinguish between saturated and unsaturated fats. The smallest unsaturated fatty acids are acetic, propionic, butanoic (often called butyric), pentanoic (valeric) and hexanoic acids. They are also called short chain fatty acids. They contain 2, 3, 4, 5 and 6 carbons, respectively. Some of them are produced by fermenting bacteria in the gut. Medium-chain fatty acids have 7-12 carbons, while long chain fatty acids have over 12 carbons. Stearic acid has 18 carbons that are arranged in a nearly linear or zig-zag configuration. The long chain alkyl groups of fatty acyls can pack closely together in cell membranes, making them more rigid and they have higher melting points. So, animal fat is solid at room temperature. Unsaturated fats are not healthy. Their consumption should be limited in a healthy diet [1].

Since stearic acid has no C=C double bonds, it can be abbreviated as 18:0. In the official IUPAC system of nomenclature, the carbon in the COOH group is carbon number 1. It has a long alkyl ($C_{17}H_{35}$) chain attached to it with 16 $–CH_2–$ groups and a terminal $–CH_3$. This is called the omega (ω) carbon. Note that the hydrogens are not shown in the $–CH_3$ and $–CH_2–$ groups, which are just represented as points connected by line segments.

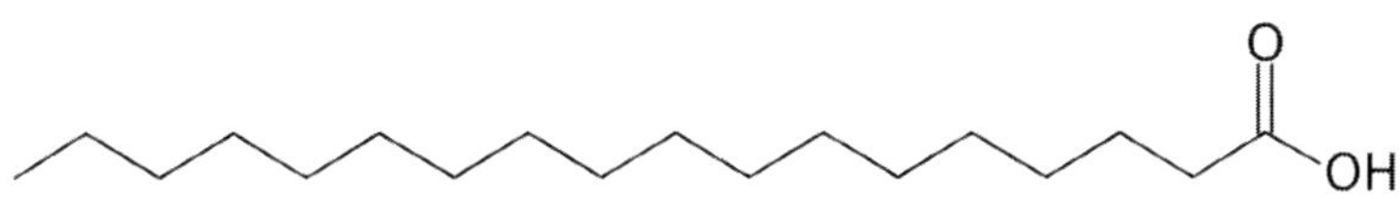


Figure 3. Stearic acid, 18:0.

When a molecule of stearic acid reacts with glycerol, a monoacylglycerol or a diacylglycerol, it becomes a stearoyl, as shown in Figure 4.

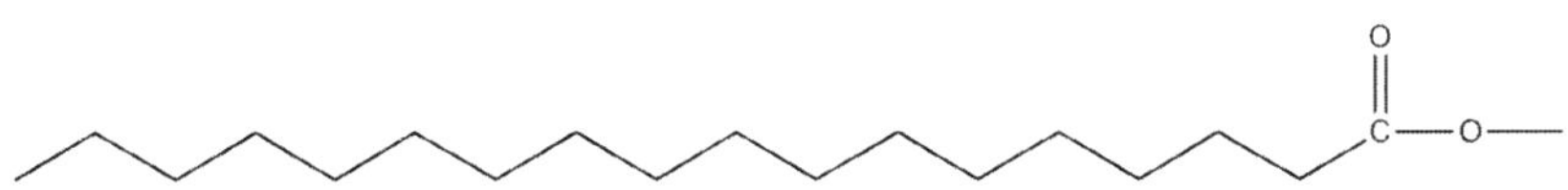


Figure 4. Stearoyl, 18:0.

If a fatty acid or fatty acyl contains one or more C=C bonds, it is unsaturated. Oleic acid is an 18-carbon fatty acid with one C=C bond, so it is abbreviated as 18:1. For simplicity, its structure is sometimes drawn as shown in Figure 5.

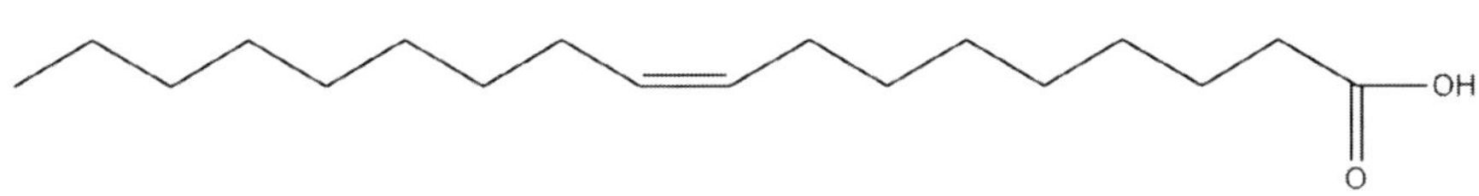


Figure 5. Oleic acid, 18:1, drawn in a linear form, for simplicity.

In fact, it is quite non-linear, as shown in Figure 6.

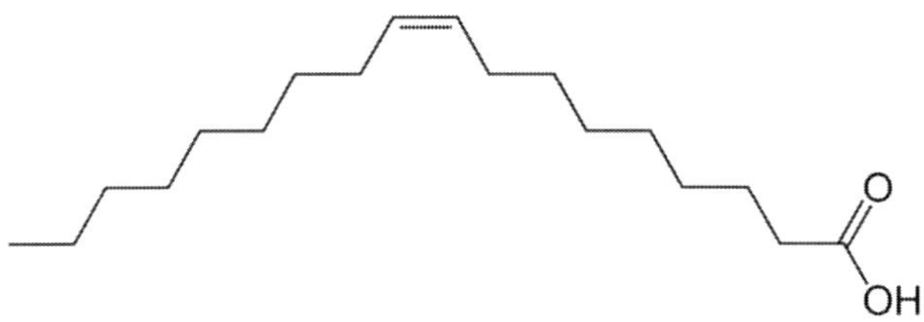


Figure 6. Oleic acid, 18:1, a *cis* fatty acid, drawn in its nearly genuine non-linear form.

As in the other Figures, neither the $-CH_3$, $-CH_2-$ or -HC=CH- groups are not shown explicitly. In any HC=CH bond, the hydrogens can be either on the same side of the C=C bond as they are in oleic acid, or they can be on opposite sides, as in *trans* vaccenic acid, as shown in Figure 7.

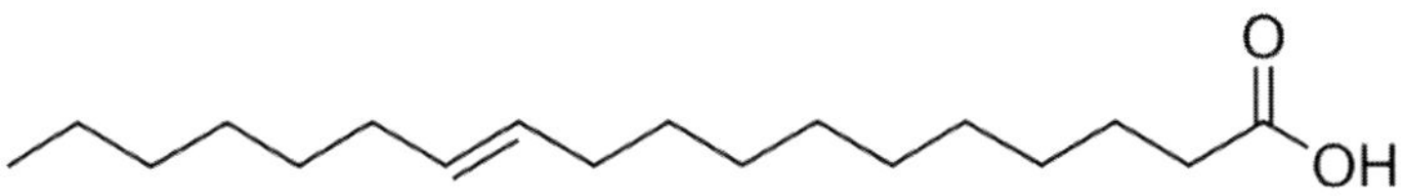


Figure 7. *Trans* vaccenic acid, 18:1.

That is, when the hydrogens are on the opposite side of the C=C bond, they are in the *trans* configuration. Unlike the *cis* fatty acids and fatty acyls, their long chain alkyl groups are nearly linear and can pack together closely, giving them higher melting points and making them unhealthy [1]. The main dietary sources of *trans* fats are partially hydrogenated vegetable oils that are used to make fried foods and many processed foods [1].

Fatty acids and fatty acyls can have more than one C=C bond. For example, linoleic acid (18:2) is an 18-carbon fatty acid with two *cis* C=C double bonds, as shown in Figure 8. Because it has a C=C six carbons from the terminal ω carbon, it is a member of the ω-6 fatty acid family.

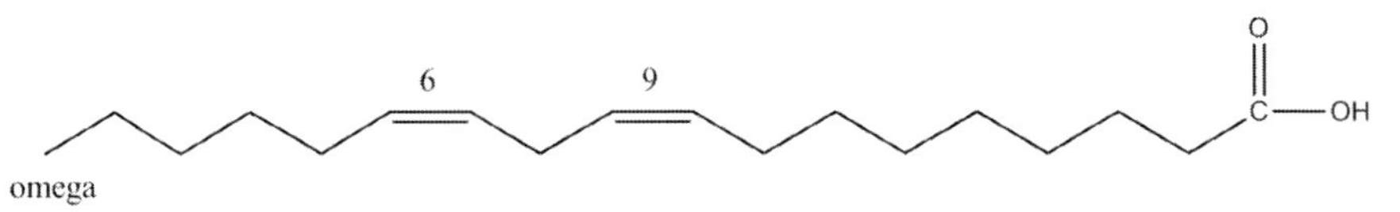


Figure 8. Linoleic acid, 18:2, an ω-6 *cis* fatty acid.

There is also a fat called alpha-linolenic acid, or α-linolenic acid, 18:3. Since it has a C=C bond three carbons from the terminal ω carbon, it is an ω-3 fat. Like other ω-3 fats, it is an important part of a healthy diet.

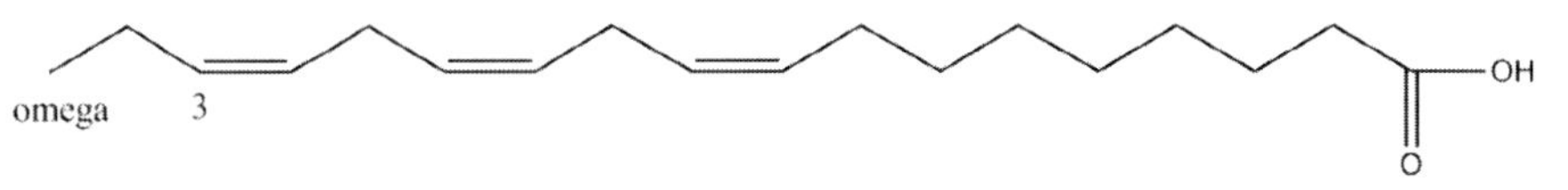


Figure 9. α-linolenic acid, 18:3, an ω-3 *cis* fatty acid.

Fatty acids and fatty acyls can also have a mixture of *cis* and *trans* C=C bonds, as in catalpic acid, another type of 18:3, shown in Figure 10.

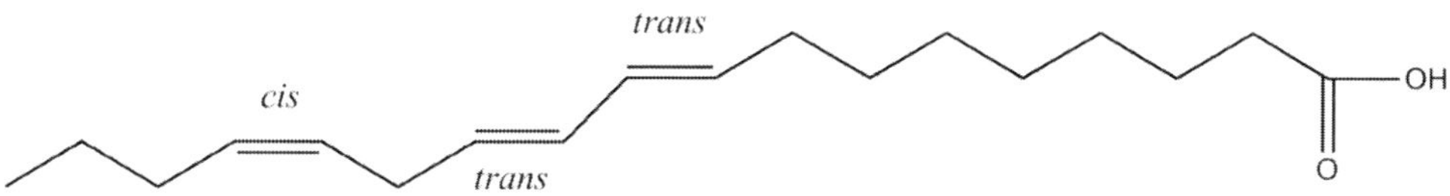


Figure 10. Catalpic acid, 18:3, with one *cis* and two *trans* C=C bonds.

In contrast, the very healthy punicic acid that is found in pomegranates has two *cis* and one *trans* C=C bond, as shown in Figure 11.

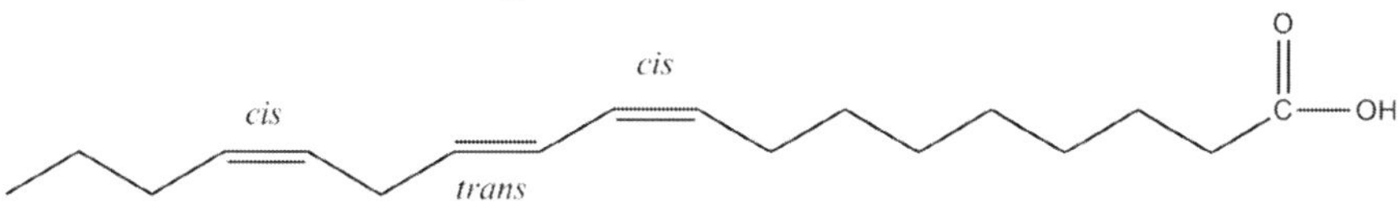


Figure 11. Punicic acid, 18:3, with two *cis* and one *trans* C=C bonds.

To avoid obesity, metabolic syndrome, smoldering inflammation, diabetes, stroke, heart disease and cancer, it is important to eat the right types of fat (unsaturated and omega-3 fats) and avoid over-consuming the wrong kinds, such as saturated and *trans* fats, as well as cholesterol [1]. Unsaturated and omega-3 fats are also important in helping to resolve inflammation and protect against it. They are metabolized into resolvins, protectins and lipoxins. Lipoxins are biosynthesized from the polyunsaturated fatty acid called arachidonic acid, shown in Figure 12.

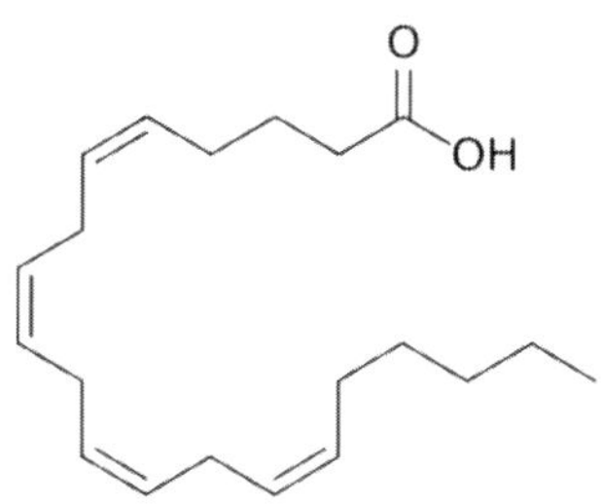


Figure 12. Arahcidonic acid, a precursor of lipoxins.

Protectins and resolvins are made from omega-3 fats, such as α-linolenic acid (ALA), eicosapentaenoic acid (EPA) and docosahexaenoic acid (DHA) [1]. "Omega-3 fats can be taken as dietary supplements such as fish oil and flaxseed oil. They are also present in fatty fish, such as salmon. Eskimos eat much fatty fish and they have a very low incidence of arthritis and heart disease" [1].

There is also a hormone (or adipokine) produced by adipose cells that is called adiponectin. It increases insulin sensitivity, decreases glucose production in the liver and decreases glucose uptake in the muscles [1]. "All of these are antidiabetic activities. Adiponectin is abundant in human blood (5–20 μg/ml), and its plasma concentration decreases with fat accumulation in the body. Plasma adiponectin concentrations are lower in patients with diabetes and ischemic heart disease. It activates glucose uptake in muscles, enhances β-oxidation of fatty acids, and suppresses glucose production in the liver. Adiponectin receptors are found in skeletal muscle and the liver and they have important roles in the regulation of glucose and lipid metabolism, along with inflammation and oxidative stress. Adiponectin also decreases the concentration of lipids and cholesterol in the circulating blood and is anti-inflammatory" [1].

"Excess body fat (especially abdominal fat) is associated with dysregulated energy metabolism and less adiponectin secretion [1]. "Excess body fat also increases the production of a pro-inflammatory molecule called tumor necrosis factor alpha, or TNF-α. Adipose cells (adipocytes) can become dysfunctional because of inflammation and improper regulation of crucial enzymes. So, adipocytes are not just storage cells for triglycerides, but complex cells that must be properly regulated" [1]. Adipose tissue is highly vascularized [5]. "Several crucial pro-angiogenic proteins, including leptin, adiponectin, hepatocyte growth factor 1 (HGF-1), angiopoietin-2, and vascular endothelial growth factor A (VEGF-A) are secreted by adipocytes. When VEGF-A is up-regulated in adipocytes, it improves vascularization and causes a "browning" of white adipose tissue, with an increase in energy expenditure and resistance to high fat diet-mediated metabolic insults [5].

However, the effects of angiogenesis inhibitors depend on the biological context [1]. "Angiogenesis facilitates healthy fat expansion. Adipocytes have a smaller average size" [1]. There is an absence of hypoxia, minimal fibrosis, and virtually none of the hallmarks of inflammation that are characteristic of dysfunctional adipose tissue [5]. VEGF-A also causes white adipose tissue to assume a phenotype resembling more closely brown adipose tissue [5]. So, dietary angiogenesis inhibitors in mushrooms and onions may have a negative health effect when gaining weight, but are probably helpful when one's weight is stable, even if obese [1]. When an adult is gaining weight, inhibiting angiogenesis might decrease healthy fat pad expansion, but once one's weight is stabilized, inhibiting angiogenesis might inhibit this critical step in cancer metastasis [1].

So, fats and adipose cells are important, as are dietary sources of healthy unsaturated fats, such as punicic acid in pomegranate seeds and the oil that is made from them.

Another major class of bioactive compounds in pomegranates and products made from them are phenolic acids. Like fatty acids, they are weakly acidic and are more soluble in water when the pH is high enough to form the anionic carboxylate ($COO^-$) ion. They can also form phenolate anions at pH>8 [1]. Unlike fatty acyls in dietary triglycerides, they are relatively hydrophilic. Triglycerides dissolve in hexane and chloroform, while phenolic compounds can dissolve in methanol and ethanol as well as mixtures of them with water. The structure of phenol was shown in Figure 4 of Chapter 1. It has a single, slightly acidic –OH group attached to a benzene ring. There are also phenolic acids that have both an acidic –OH and an acidic –COOH. One of the most important ones in pomegranates is gallic acid, as shown in Figure 13.

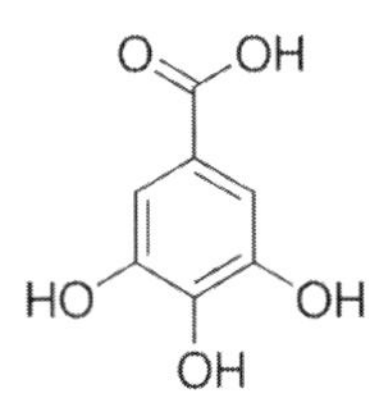


Figure 13. Structure of gallic acid, which is found in pomegranates and is used as a standard when quantifying total phenolic compounds.

There are also phenolic compounds that do not have a carboxylic acid group. This includes ellagic acid (Figure 14), one of the most important phenolic compounds found in pomegranates.

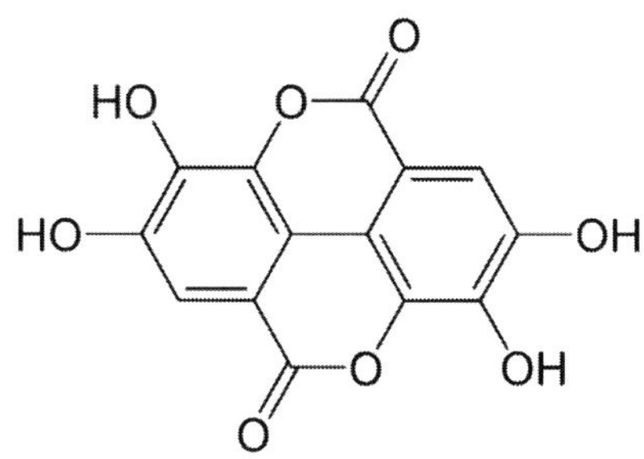


Figure 14. Ellagic acid.

Ellagic acid is an example of a flavone. The flavone backbone is shown in Figure 15.

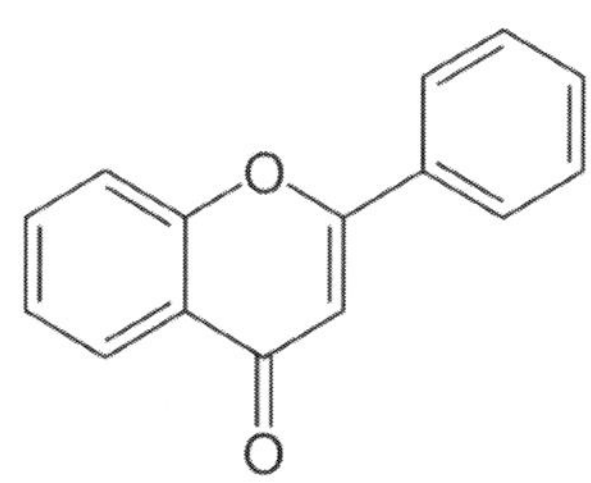


Figure 15. Flavone Backbone.

Anthocyanins make up an important class of flavones that are found in pomegranates. The generalized structure of an anthocyanin is shown in Figure 16.

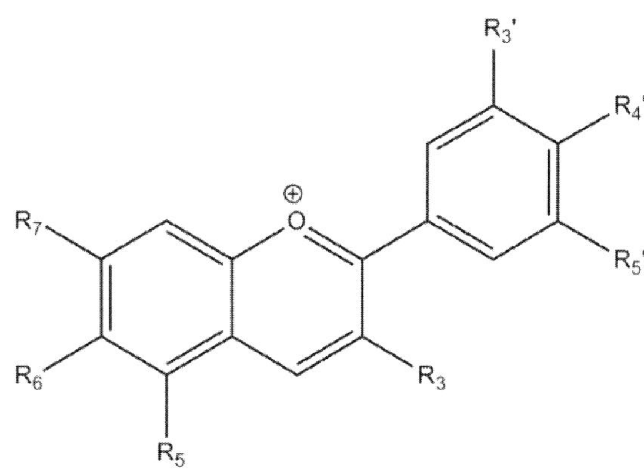


Figure 16. Anthocyanin generalized structure. The R groups can be –H, -OH or –$OCH_3$.

The structure of cyanidin, an important anthocyanin in pomegranates, is shown in Figure 17.

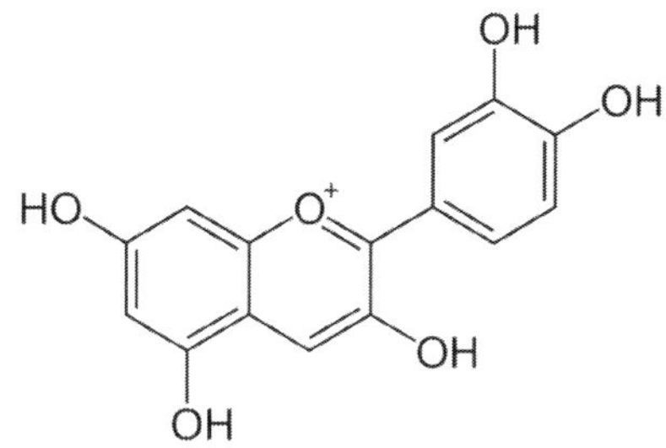


Figure 17. Cyanidin.

Cyanidin, like other anthocyanins, can be glycosylated to make cyanidin 3-O-glucoside, cyaniding 3-O-rutinoside and other compounds.

Tannic acid is another important phenolic compound in pomegranates. Its structure is shown in Figure 18.

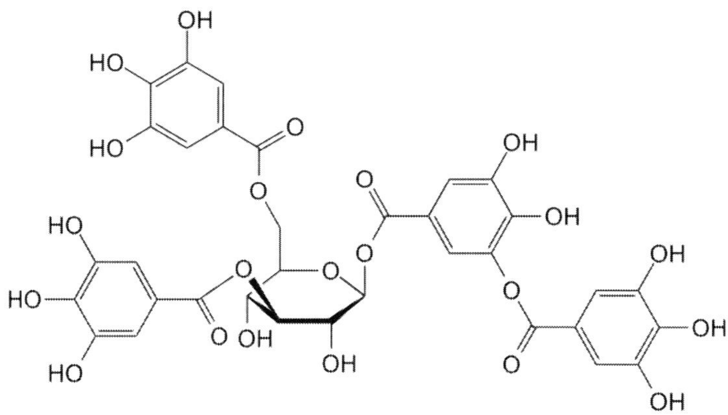


Figure 18. Tannic acid, a type of tannin.

Punicalagin is an important phenolic characteristic of pomegranates with an even more complicated structure, as shown in Figure 19.

Terpenes are also in pomegranates. The structure of one of them, oleanoic acid, was shown in Figure 1 of Chapter 4. Ursolic acid is another. Its structure is shown in Figure 20.

Different analytical methods were mentioned in the chapters on chemical composition and health effects. They include gas chromatography (GC), liquid chromatography (LC), high performance LC (HPLC), ultra-performance LC (UPLC) and mass spectrometry (MS), combined with GC (GC-MS) or LC (LC-MS). These will be described next.

For a compound to be analyzed by GC or GC-MS, it must be volatile or convertible to a gas after undergoing a suitable reaction (derivatization). Either the original compound or its volatile derivative can be identified by comparing the mass spectrum to that of a standard or a computer database of standards. The National Institute of Standards and Technology (NIST) has such a database that was used to identify volatile compounds in pomegranates that affect its aroma [6]. The mass spectra are produced by the relatively strong ionization technique used in GC [1, 6]. However, GC is more often used to determine the fatty acyl composition of triglycerides in the oil [7]. Unfortunately, triglycerides are not volatile. So, before they can be analyzed by GC they must be saponified (hydrolyzed) to make glycerol plus three fatty acids.

The fatty acids are then reacted with activated methanol (in the form of $BF_3$-$CH_3OH$) to make fatty acid methyl esters (FAMEs), which are volatile and quantifiable by GC.

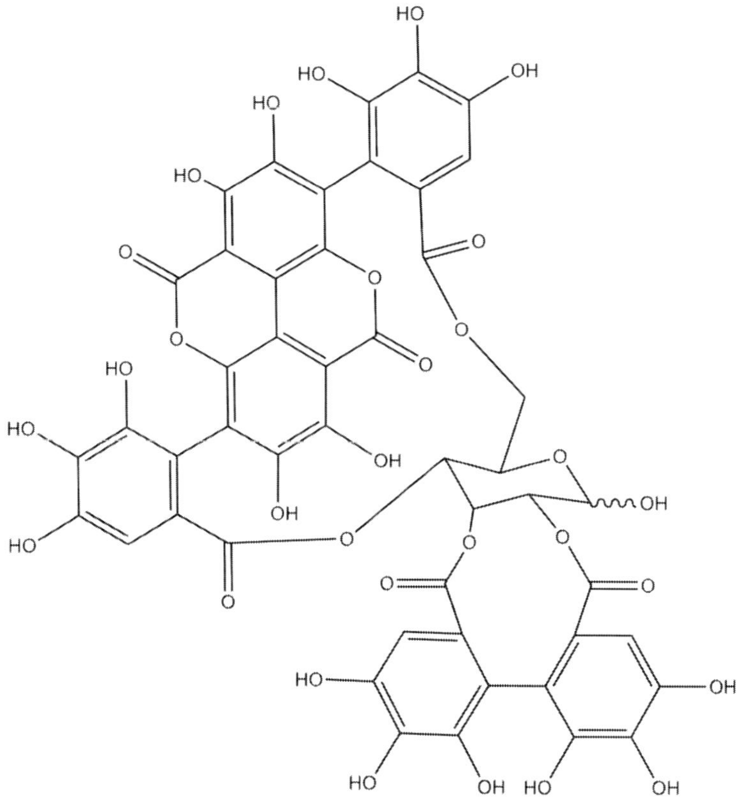

Figure 19. Punicalagin.

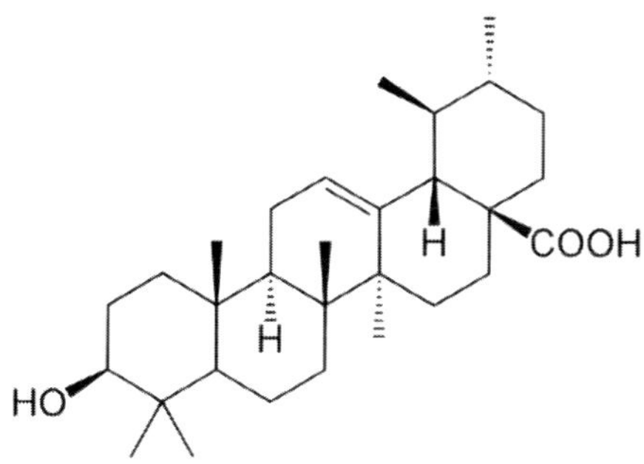

Figure 20. Ursolic acid.

Nonvolatile phenolic compounds and amino acids are analyzed better by LC. Like GC, LC uses a column to separate compounds in a mixture when analyzing it. Unlike GC, LC can be used not just for analysis, but also for preparative scale separations. LC without a packed column was used to separate large amounts of gallic acid and punicalagin from peels [8]. High-speed countercurrent chromatography (HSCCC) was used for preparative scale separations [8]. It used two immiscible phases. One is the mobile phase and the other is the stationary phase (which is kept stationary by centrifugal force). The mobile phase was pumped at 2 mL/min, while the stationary phase rotated at 800 rpm [8]. Effluents coming off the HSCC were analyzed by UPLC for separation and electrospray MS (ES/MS) for detection [8]. UPLC, like HPLC and LC-MS, uses a packed column to separate components or analytes. Some columns are disposable and are used for sample clean-up before analysis, especially when analyzing foods for pesticides [9]. They are used for solid phase extraction (SPE) of interfering compounds before doing an analysis [9]. They operate under low pressure. Analytical columns used in many older papers used HPLC columns that could take pressures up to about 2000 psi (pounds per square inch) without leaking. Such papers often used UV-visible absorbance for detection. The oldest papers sometimes used UV detectors that operated at one or two fixed wavelengths. More modern UV detectors have an array of photodiodes so the entire UV (and visible) spectrum can be detected at once. UV detectors are relatively cheap and easy to use, but not all analytes absorb UV light and the ones that do have different absorptivities. So, calibration curves must be made for each analyte to do quantitative analysis. Also, compounds with very similar structures may not be separated on an HPLC or UPLC column and may not be distinguished by their UV spectra. So, false positives can occur.

UPLC columns can take much higher pressures that HPLC columns and are packed with smaller particles, which improves the separation of analytes. They also require less mobile phase and are ideal when using MS detection. It is more expensive, but MS can detect many more compounds than UV. When operated at high resolution, MS can identify an analyte. Also, isomers that produce the same UV spectrum can often produce different mass spectra once the primary ion is fragmented.

That is, MS measures the masses of ions. So, one part of a mass spectrometer will produce ions, another will separate them and another will detect them. In LC, relatively soft ionization techniques are used, so fragments are not always produced, just the molecular ion. Electrospray (ES) is one way to generate ions for LC-MS. It can be used in either the positive ($ESI^+$) or negative ($ESI^-$) mode. That is, negative ions can be produced from compounds like phenols by removing one hydrogen (H) to produce a molecular ion minus one H, or $[M-H]^-$. On the other hand, ES+ is used to ionize compounds like amines that readily accept a hydrogen to make $[M+H]^+$. So, in $ESI^+$ the mass of the $[M+H]^+$ is one Dalton higher and in $ESI^-$ $[M-H]^-$ is one Dalton lower than the molecular weight of the compound being analyzed (the analytes).

Another soft ionization technique is matrix assisted laser desorption and analysis, or MALDI. In it, a chemical is added to the sample matrix to make it ionizable when laser light strikes it. It causes the analyte to be desorbed and ionized so that a mass spectrometer can detect it. Usually, a time-of-flight (TOF) mass spectrometer is used instead of a column to separate the analytes. They are separated based on the amount of time it takes for them to travel through the flight tube and reach the MS detector. That, in turn, is based on the size or the molecular weight of the analyte. So, MALDI-TOF MS is an excellent technique for analyzing many compounds, but it can't distinguish between isomers that have the same molecular weight.

Many of these bioactive compounds control the expression of genes that are translated into messenger RNA (mRNA), which are then translated into proteins. The genes are interspersed along the chromosomes of plants and animals. In pomegranates and humans, the chromosomes are almost always in pairs (diploid), although some healthy human cells can be polyploid and considerable chromosomal abnormalities emerge in cancer. Somatic cells can go through cycles. Most often, they are in a resting or gap ($G_0$) phase. Still, they can divide during interphase to produce two genetically identical cells in mitosis. This occurs in steps or phases, from gap or growth 1 ($G_1$), to synthesis (S), to gap or growth 2 (G2), to final cell division (mitosis). Some cells can also go through reductive division, or meiosis, to produce gametes (sex cells). There are checkpoints at $G_1$ and $G_2$ to ensure that the cell is truly ready (free from DNA damage) for the next step in the cycle. In sexual reproduction, plant and animal cells undergo reductive cell division, or meiosis, to produce four genetically unique cells in the case of diploids like humans and pomegranates. As mentioned in Chapter 2, cytogenetic analysis of pomegranates can be done when cells are undergoing meiosis, which occurs in two stages, I and II. During meiosis I, homologous chromosomes are separated to produce two haploid cells. This is followed by Prophase I, in which DNA can be exchanged between homologous chromosomes by recombination. There are several steps in Prophase I, including diplotene, in which homologous chromosomes start to separate and in which non-sister chromosomes can exchange genetic material as they form DNA crossover points called chiasma. Finally, plants (unlike humans) can have accessory or B chromosomes, which useful in analyzing pomegranate genetic diversity, as mentioned in Chapter 2.

In the $20^{th}$ century it was thought that DNA was the "blueprint of life" because it supposedly had all the instructions needed to encode all the proteins needed for the "machinery of life" [1]. "This was part of reductionist thinking, in which many diseases were thought to be caused by the defect in a single gene or protein that was coded for by that gene. Moreover, humans were considered to be like machines, created to spread our selfish genes. Much of human (and animal) behavior was thought to be under the control of our selfish genes" [1]. In his book, The Selfish Gene, Dawkins wrote that "genes swarm in huge colonies, safe inside gigantic lumbering robots sealed off from the rest of the outside world ... They are in you and me; they created us, body and mind; and their preservation is the ultimate rationale for our existence" [10].

Yet, there is an opposite metaphor about genes that is more consistent with systems thinking [1] that says that genes "are trapped inside huge colonies, locked inside highly intelligent beings, molded by the outside world, communicating with it by complex processes, through which, blindly, as if by magic, function emerges. They are in you and me; we are the system that allows their code to be read; and their preservation is totally dependent on the joy we experience in reproducing ourselves. We are the ultimate rationale for their existence" [11]. Still, this is just a metaphor, or a philosophic vision. It is useful in guiding future research and in interpreting experimental results. Metaphors are useful in teaching and communication, but are not scientific facts. So, it is probably impossible to think of any experiment that can "prove" that either of these contradictory metaphors is correct [11]. Perhaps a more useful metaphor for the genome is that it is like a CD or flash drive that contains a database that by itself, does not code for anything [11].

Instead of a binary database, with ones and zeroes, like a CD, the genome is a quaternary database, with four monomers, or letters (A, T, G and C) instead of two numbers [1, 11]. "Also, different parts of the genome database that are called genes can have very different meanings (or functions) depending on which cell or organism is accessing (or reading) the database. Genes can 'code' for very different proteins and types of RNA in different cells and different organisms. Genes are usually not 'selfish'. They usually do not determine the fate or behavior of the organism. In fact, the roles of many genes are affected by the environment of the cell, which is affected by the proteins (and other molecules and ions) that enter the cell from the environment. For example, queen bees and female worker bees have the same genome, but the queen bees grow much larger, live much longer, produce eggs and have very different behavior than the workers. The workers produce royal jelly that the queen eats, causing her genes to 'behave' very different than the same genes in the genomes of workers, who don't eat royal jelly" [1].

Moreover, there are mobile genetic elements, also known as jumping genes and transposons [1]. "They are "selfish" because their only function seems to be to make more copies of themselves, with no benefit to the organism (including humans) in which they are found. For example, 10.6% of the total human DNA contains repetitive sequences, called *Alu* elements, which are short interspersed elements (SINES) containing 100-400 base pairs long. There are also long interspersed elements (LINES), which are up to 9000 base pairs long. The human genome contains about 868,000 LINES and they are 20.4% of the total genome. Transposons can be mutagens if they are inserted into a functional gene. So, there is a big difference in the usage of the term "selfish genes". Richard Dawkin's book used the term as a metaphor, while others use the term to indicate an actual scientific concept concerning transposons and a few genes" [1].

Also, our concept of genes, selfishness and cooperativity continues to evolve [1]. "Genes can be made of DNA or RNA. There is a division of labor in cells. DNA and RNA are used to store and process genetic information, while proteins, lipids and carbohydrates primarily perform regulatory, metabolic and structural roles. Some proteins also help control the expression of genes and work with self-assembled ribosomal RNAs to make themselves and other proteins in ribosomes. Even though the type of proteins called enzymes catalyze almost all the reactions in cells, some reactions can be catalyzed by catalytic RNAs, also known as ribozymes" [1]. One ribozyme, from the bacterium *Azoarcus* can assemble itself after being fragmented [12]. Researchers have shown that similar RNA fragments can self-assemble and form cooperative hypercycles [13]. Through such cooperative hypercycles, such RNAs were able to form ribozymes, which probably pre-dated DNA or proteins. So, the original genes are thought to have been cooperative, not selfish, and made from RNA, not DNA [12, 13].

Not all DNA codes for mRNA or its precursor, hnRNA [1]. "Some code for small interfering RNAs (siRNAs), which ensure genomic stability by silencing endogenous selfish genetic elements such as retrotransposons and repetitive sequences. Other portions of DNA can be translated into precursors of ribosomal RNA (rRNA) and transfer RNA (tRNA). There are also smaller forms of RNA, including small interfering (siRNA), micro (miRNA) and Piwi-interacting (piRNA), which can prevent the translation of mRNA into proteins. In plants there are small, mobile RNAs that can move across cell membranes and silence the transcription of genes in other plant cells" [1].

Like hnRNA, precursors of rRNA and tRNA are chopped up (hydrolyzed) and put back together (ligated) to make mRNA, rRNA or tRNA [1]. "Sometimes, the precursor RNA molecules act as their own catalysts, accelerating the process. So, not just mRNA, but also other forms of RNA are important. In fact, rRNA, which is produced in the nucleolus in eukaryotic cells, accounts for most of the RNA in eukaryotic cells. Ribosomes are assembled in the nucleolus. Ribosomes are the place where proteins are made from the mRNA template and aminoacyl tRNAs that bring each amino acid to the growing polypeptide chain. Also, other forms of RNA can regulate gene expression and the translation of mRNA into many different kinds of proteins. For example, miRNAs are coded for by the genome and are processed into smaller pieces (22 base pairs) that can interact with homologous regions on mRNA molecules to modulate their translation into proteins. A much longer RNA molecule (16,000 nucleotides) called XIST binds to one of the two X chromosomes in female mammalian diploid cells and inactivates it. There are also ribozymes that can catalyze their own hydrolysis (self-cleaving ribozymes) and are converted to a form that can control the expression (transcription) of genes" [1].

So, to make human proteins, first the DNA is transcribed into hnRNA, which is converted to mRNA, which can be translated into polypeptides or proteins, which provide structural support, do much of the work in a cell, and aid in cellular communication. Many proteins catalyze biochemical reactions [1]. "These are called enzymes. Some of them exist in different forms called isozymes that catalyze the same reaction, but often in different cells and/or different biological environments. When RNA molecules act as catalysts, they are called ribozymes, RNA enzymes or catalytic RNA. On the other hand, bacterial DNA is transcribed directly into mRNA (without any hnRNA intermediate). The central dogma of molecular biology in the 20$^{th}$ century was that information could not flow from protein to RNA to DNA and that DNA codes for mRNA, which is translated into proteins. Genes were said to code for proteins. As mentioned earlier, there was even a hypothesis called the one gene one polypeptide hypothesis, which said that each gene codes for a single polypeptide. Now, scientists know that there are important exceptions to this. Millions ($>10^6$) of different types of proteins are made in humans from only about $2x10^4$ genes. This is done by splicing portions of gene segments, such as the variable, diversity and joining regions of genes that code for antibodies in the human immune system. Also, some genes code for small pieces of RNA, such as miRNA, rather than proteins. Some of the miRNA can bind to mRNA and affect its translation into proteins. Also, miRNA plays a role in chromosome segregation, differentiation of developing cells, metabolism and programmed cell death (apoptosis). Defects in miRNA have been implicated in cancer and diabetes. For example, miRNA-802 is upregulated in type II diabetes. Also, double-stranded miRNA can bind to complementary portions of DNA, turning off its transcription into mRNA. Other types of RNA, called noncoding RNA-activating (ncRNA-A), can activate transcription" [1].

There are also small inhibitory RNAs (siRNAs) that contain 19-21 nucleotides [1]. "They can be used in the laboratory to knockdown specific proteins. In the last few years, it has been found that some pieces of DNA code for other types of RNA molecules, including some that are catalysts (like enzymes) and others that regulate transcription and translation (such as small inhibitory, or siRNA). Some of these RNA molecules can bind to pieces of DNA or mRNA, turning them off or on. During the life of a cell, different pieces of DNA need to be turned on or off at just the right times, in response to environmental stimuli. The RNA and proteins that are needed are made at the right times and in the right locations in the cell. On the other hand, in cancer, transcription of some of the DNA (oncogenes) can be left on when it should have been turned off" [1].

Still, the so-called central dogma still tells only part of the story [1]. "Some parts of DNA are used as templates to help produce hnRNA and then single-stranded mRNA, which codes for proteins or polypeptides. The piece of DNA that codes for a polypeptide or protein is called a gene. Other pieces of DNA code for types of RNA molecules that have functions that are so important that most scientists use the term RNA genes for them, too. So, the genotype of an organism is the collection of genes that exists in that organism. However, the genes are expressed in different ways in different cells and in different people and at different times in their lives. The outward appearance of the cell or the organism that results from the selective expression of genes is called the phenotype. However, the phenotype can be affected by the environment, regardless of the genotype" [1]. For example, the sex of many reptiles is determined by the temperature during embryonic development [14]. When eggs of the Australian central bearded dragon lizard (*Pogona vitticeps*) were incubated at relatively high temperatures, genotypic males (with ZZ genotype) were converted to phenotypic females [14].

So, let us see how genes are expressed in human and other eukaryotic cells [1]. "As mentioned earlier, the gene codes first for hnRNA, which is chopped up (hydrolyzed) into pieces to remove insertion elements, which are called introns. The other parts (exons) are then re-combined (ligated) to make mRNA, which moves out of the cell nucleus and into the ribosome, which contains tightly bound water, ions, proteins and RNA. The ribosome contains a complex of ribosomal RNA (rRNA) and proteins. The ribosome catalyzes the polymerization of amino acid monomers to make proteins using mRNA as the template. It takes three ribonucleic acids to code for each amino acid. This is called a triplet code. For example, if a piece of DNA has the following sequence of bases: ATGCAT, it would be transcribed into this piece of mRNA: UACGUA. The UAC would be translated into the amino acid tyrosine and the GUA would be translated into valine" [1].

One amino acid at a time is brought to the mRNA [1]. "Each amino acid is attached to a molecule of transfer RNA, or tRNA. The amino acids combine to form polypeptides and/or proteins. Remember, DNA is made of deoxyribose, phosphate and the four bases, adenine (A), guanine (G), thymine (T) and cytidine (C)."RNA is made from ribose and three of the same bases used in DNA, A, G and C. Instead of T, it uses uracil, abbreviated by U. When a base has a sugar attached to it, it is called a nucleoside. In DNA and RNA, the base is attached to a sugar. In RNA, the sugar is ribose, in DNA it is deoxyribose. So, the letter A stands for adenosine in RNA, and dA for deoxyadenosine in DNA. G stands for guanosine and dG for deoxyguanosine, dT stands for deoxythymosine. U stands for uridine, and dC stands for deoxycytosine. However, many people just use A, T, G and C for the deoxy nucleotides, not dA, dT, dG or dC. Usually people say that A-T and G-C are the base pairs in DNA (not dA-dT and dG-dC)" [1].

In RNA or DNA, the nucleosides have a phosphate group that links the sugars together with covalent bonds [1]. "Thus the letter p in CpG stands for the phosphate that connects them with covalent bonds. Phosphorylated nucleosides are called nucleotides. To make DNA or RNA, one needs the nucleoside triphosphates. RNA is made by reacting ATP, UTP, GTP and CTP with each other. DNA is made by reacting dATP, dTTP, dGTP and dCTP, where the d stands for deoxy. In the RNA polymerase-catalyzed reaction between two nucleoside triphosphates, five of the six phosphates are byproducts of the reaction, leaving just one phosphate between each nucleoside. Note that at the pH of the cell, phosphates are a mixture of $H_2PO_4^-$ and $HPO_4^{2-}$" [1].

ATP is important for more than just being used to make RNA [1]. "ATP is used to store energy in cells. There are some metabolic cycles, such as glycolysis and the citric acid cycle" that produce ATP. When cells need some of this stored energy, they can hydrolyze the ATP to produce adenosine diphosphate (ADP) and/or adenosine monophosphate (AMP)" [1].

When DNA is not being copied, it exists as a double strand, in which the two strands are held together by hydrogen bonds" [1]. "Also, the DNA is wrapped around histone proteins and can become inaccessible to DNA or RNA polymerases due to supercoiling. DNA must be unwound before it can be copied into more DNA or transcribed into RNA. Epigenetic factors also affect gene transcription. When cytosine is next to guanosine (CpG) in eukaryotic genes, a methyl group is usually attached to the C. This prevents the transcription of the gene, but it does not prevent it from being copied when a cell grows and divides. So, it is not just our genes that are copied from one generation to the next, but also our epigenetics, including the degree of methylation. So, the prefix epi- means on top of and epigenetics is a layer of control on top of the classical DNA-based genetic control" [1].

There are different levels of organization in human cells [1]. "Pieces of DNA may be very close to each other or even adjacent to each other. A modular DNA sequence that encodes one or more proteins that perform a similar function is called a gene cassette. Some proteins are able to bind to one piece of DNA and either activate or inactivate the transcription of nearby DNA, especially in bacteria. So, cause and effect work both ways. Not only can DNA help cause RNA and proteins to be made, but RNA and proteins can also help cause DNA to be transcribed, or prevent it from being transcribed, or induce transcription into different mRNA molecules, which are translated into different proteins. As mentioned earlier, epigenetic controls are important, too. Finally, some pieces of RNA can bind to DNA or mRNA to enhance or inactivate translation or transcription. This happens when one of the two X chromosomes in females is inactivated to produce a Barr body. Other types of RNA (miRNA) can bind to mRNA to keep it from being translated into protein [1].

In human and other types of eukaryotic cells, genes are seldom found in unbroken stretches of DNA [1]. "That is, DNA is first transcribed into hnRNA, which contains insertion elements (introns) that are removed before the remaining exons can be spliced together to make mRNA. The RNA can be spliced in different ways to make different proteins and regulatory RNA molecules, depending on the cellular environment and the needs of the cell. Also, gene segments can be located on different chromosomes or different regions of the same chromosome. Genes for the constant, diversity and joining regions of the type of proteins called antibodies can be mixed and matched in many different ways to make millions of different antibodies" [1].

Many of our cells are being continuously broken down and re-made. However, there are also stem cells which are capable of making copies of themselves, and they can differentiate into new cells with new properties. This requires that the appropriate genes are turned on or off at the right times. Enzymes called DNA demethylases catalyze the removal of methyl groups from cytosines, thus activating genes. Almost all the somatic cells in the human body (except red blood cells) have the same 23 pairs of chromosomes, but they are in different environments, which affect epigenetics. As mentioned earlier, the classical definition of a gene is that it codes for a protein. The sequence of bases in DNA is called the genetic code [1]. At one time, scientists thought that this code was like a blueprint. They thought that it was a fixed set of instructions that could not be cut up and rearranged. One gene was thought to code for one protein, and genes had fixed locations on chromosomes.

However, in 1950, Barbara McClintock published a paper that showed data indicating that genes could move around, or jump [1, 15]. "This paper was ignored by men for many years, but we now know that there are jumping genes, more commonly known as transposable elements, transposons or mobile genetic elements. For example, when some highly differentiated white blood cells make proteins called antibodies, they have pieces of genes that rearrange themselves in the chromosomes before coding for the antibody. Transposons can also cause mutations and change the DNA in genome. The most common transposon family in humans is called Alu, which is a family of repetitive bases, about 300 base pairs long. Most transposons in human cells copy themselves through RNA intermediates and are called retrotransposons" [1].

We also have learned that some proteins can be chopped (hydrolyzed) into smaller peptides, each with different properties [1]. "We have also learned that many proteins are modified by attaching lipids, sugars and/or phosphates after they are made. Those reactions are catalyzed by enzymes. There are other proteins that help a newly synthesized protein fold into the proper shape in healthy cells. When proteins fold into the wrong shape, they can cause Alzheimer's disease. In healthy cells, defective proteins can be marked for hydrolysis by attaching small proteins called ubiquitins (molecular weight ~8500). They are ubiquitously expressed and highly conserved from yeasts to humans. Moreover, ubiquitin can be attached to to some proteins and cause a variety of other effects. They can cause proteins to be degraded in proteasomes or just modulate protein activity. Ubiquitin is a node in a network that regulates biological processes and propagates information in human cells. So, it is important to remember that DNA, RNA, proteins, lipids and sugars interact in the complex network that is the web of life" [1].

It is now known that genes can transfer horizontally as well as vertically [1]. "Vertical transfer is when genes are transferred from parents to their children. Horizontal transfer is when genes are transferred from one organism to another. This is how genes are transferred in bacteria. It can also occur in unicellular eukaryotes and in some multicellular eukaryotes, called metazoans. Genes can be transferred from bacteria into human cells. So, humans and other organisms can be exposed to antibiotic-resistant bacteria even if they have never been given an antibiotic. The genes that code for antibiotic resistance can be passed from one species or type of bacteria to another, in horizontal transfer. Also, about 8% of human DNA appears to have come from viruses, instead of our vertebrate ancestors" [1].

So, some genes code for proteins [1]. "Unlike DNA and RNA, polypeptides and proteins are made from amino acids. The amino acids react with each other, splitting off water and forming amino acyls, similar to fatty acids reacting with glycerol to form fatty acyls as part of triglycerides. Proteins can provide essential structure to cells (tau protein), restrict access to DNA (histones), activate or repress DNA transcription (CCAAT enhancer-binding protein), act as hormones (insulin) and catalyze biochemical reactions (DNA polymerases). They are involved in just about every biological process in living systems. When a protein is made, it folds into specific shapes. The primary structure of a protein is the amino acid sequence. The amino acids are held together in covalent peptide bonds that are not easily broken. Some portions of a protein can fold into structures called an α-helix and a β-sheet. These are held together by relatively weak hydrogen bonds. The α- helices and β-sheets can also fold into more complex tertiary structures. Parts of some proteins can be held together by disulfide bonds, formed between two cysteine residues. For example, insulin is made up of two polypeptide chains containing 21 and 30 amino acids. The chains are held together by four disulfide bonds. When two or more proteins and/or their subunits interact with each other, they form even more complex structures, called quaternary structures. For example, four identical subunits of hemoglobin interact with each other, forming the quaternary structure of the tetrameric hemoglobin.

Other proteins are important in regulating the transcription of genes. Some of them are membrane-bound receptors that are activated by ligands, such as free fatty acids, hormones and neurotransmitters. These ligands act as primary messengers. Their signals are transduced by other proteins, such as PI3K and protein kinase enzymes, such as MAPK. Signals are transduced through second messengers such as $IP_3$ and diacylglycerol. Some proteins act as transcription factors, such as ras, c-fos and c-jun. Some activate and others inactivate the transcription of DNA" [1].

KRAS is the oncoprotein that is most commonly activated in human cancer [1]. *RAS* is one of the most commonly mutated genes in human cancers. Oncogenes code for oncoproteins, which are upregulated in cancer. Another important oncogene is *PI3K*, which codes for the enzyme PI3K (phosphoinositide 3-kinase). The enzyme PTEN (phosphatase and tensin homolog) catalyzes the opposite reaction, so it is a tumor suppressor and the gene coding for it is downregulated in cancer. Human epidermal growth factor (hEGF, or HER), vascular endothelial growth factor, or VEGF, and the PI3K/Akt/mTOR (mammalian target of rapamycin) survival pathway are all important therapeutic target in many cancers" [1].

In addition, some viruses produce cancer by inserting (transfecting) their genes into those of the host [16]. The cancer-causing oncogenes can code for proteins that propagate signaling cascades which are not properly regulated. For example, there are viruses that cause sarcoma, or cancer of connective tissues, such as bone or muscle. One of these is called the Harvey virus and it converts the normal, healthy *RAS* gene into an oncogene (*HRAS*) and another, called the Kirsten virus converts *RAS* into *KRAS*. KRAS is the oncoprotein that is most commonly activated in human cancer [17]. KRAS can be either wild-type or mutated. If mutated, KRAS sends a signal that causes uncontrolled growth [1]. "However, oncogenic KRAS and its known downstream effectors have thus far presented intractable targets for antineoplastic drugs" [17]. Still, proteins and pathways associated with oncogenic KRAS have been identified, so they provide new therapeutic targets [17]. Each virus was named after the person who discovered it. *RAS* is one of the most commonly mutated genes in human cancers. It is found in about 20% of them, but it has proved impossible to effectively target with small molecule inhibitors. The *HRAS* and *KRAS* oncogenes code for a Ras oncoprotein, which is a mutant tyrosine kinase that is always switched on. The Ras oncoprotein affects the transcription of many genes. Several proteins are synthesized and apoptosis is inhibited [16]. *HRAS* and *KRAS* oncogenes don't do this directly, but instead act through further (downstream) pathways, including one that involves NF-κB. Hopefully, the deadly effects of *HRAS* and *KRAS* oncogenes can be stopped by inhibiting NF-κB " [1].

Ras signaling affects many cellular functions including cell proliferation, apoptosis, migration, fate specification, and differentiation [1]. The gene that codes for it is one of the most frequently mutated genes in many forms of cancer. Ras acts as a binary signal, or switch. In the resting cell, Ras is tightly bound to GDP, which is exchanged for GTP upon binding of extracellular stimuli to cell membrane receptors. In the GTP-bound form, Ras interacts specifically with effector proteins to initiate cascades of protein-protein interactions that can lead to cell proliferation. To return to the inactive state, GTP is converted to GDP-bound Ras, which is no longer able to interact with effectors. Ras can activate a number of signaling pathways, including the Raf/MEK/ERK pathway, the MEKK/SEK/JNK pathway, a PI3K/Akt/NF-κB (Nuclear Factor-Kappa B) pathway, a p120-GAP/p190-B/Rac/NF-κB pathway, and a Raf/MEKK1/IKK (I-κB Kinase)/I-κB/NF-κB pathway [18].

The Ras-Raf-MEK-ERK/MAPK pathway is involved in the control of many fundamental cellular processes that include cell proliferation, survival, differentiation, apoptosis, motility and metabolism [19]. That is, MEK is also known as MAPKK, or MAPK kinase. Also, MAPK is synonymous with ERK, so MEK is also known as ERK kinase. MAPK is mitogen-activated protein kinase, and ERK is extracellular signal-regulated kinase.

Lipids can be attached to the carboxy terminus of Ras proteins, enabling them to bind to membranes that are rich in sensory receptors. Ras proteins are arranged on the plasma cell membrane as a combination of nanoclusters and freely diffusing monomers [20]. A nanocluster contains as many as seven Ras proteins, has a radius of about 9 nm and an estimated lifetime of 0.5-1 s. Ras proteins translate extracellular signals into cell proliferation, cell survival, growth, and differentiation. Ras proteins can also be activated at intracellular membrane compartments. They often activate the ERK signaling cascade. When the *ras* gene is mutated or the Ras proteins are not regulated properly, there can be dramatic consequences. Germ line mutations that increase Ras signaling disrupt development, whereas mutational activation of Ras in somatic cells can cause cancer. There are three members of the family of Ras proteins in humans: N-Ras, HRas, and K-Ras. Germ line mutations that activate K-Ras occur in Noonan syndrome and cardiofaciocutaneous syndrome. Similar mutations in H-Ras can cause Costello syndrome and mutated N-Ras can cause autoimmune lymphoproliferative syndrome. Mutational activation in cancer can help cause several different tumors. Mutational activation of K-Ras occurs in >90% of all pancreatic cancers, N-Ras is mutated in acute myeloid leukemia and melanoma, and mutation of H-Ras is common in bladder cancer. Given the importance of Ras, cells use multiple mechanisms to potentiate and attenuate Ras activity. It becomes active in the GTP–bound state and inactive in the GDP–bound state. Guanine nucleotide exchange factors (GEFs) promote exchange of GDP for GTP to activate Ras. In contrast, GTPase–activating proteins (GAPs) promote hydrolysis of GTP on Ras to GDP, which inactivates Ras. Various posttranslational modifications (farnesylation, proteolysis, methylation, palmitoylation, and depalmitoylation) modulate the membrane affinity and localization of Ras to define exposure to downstream effectors [21].

The Ras protein can also be modified by acetylation, which activates the transport between intracellular membrane compartments of Ras proteins that can have palmitoyl (hexadecanoyl, C18:0) groups attached to them. The acylation/deacylation cycle helps control the localization of Ras and it generates a continuous Ras signal that transmits its effects between subcellular compartments [22].

Normal protein turnover mechanisms regulate the abundance of Ras [1]. "Proteasomal degradation of polyubiquitinated Ras proteins regulates their stabilities. Mono- and diubiquitination of H-Ras and N-Ras, but not K-Ras, restrict the ability to signal to ERK (extracellular signal–regulated kinase). Also, ubiquitination promotes endosomal localization and/or retention of H-Ras, which suggests that the restricted ability to signal resulted from sequestration from certain effectors. Modification of even a small percentage of total Ras proteins can lead to meaningful biological consequences" [1]. The effects of Ras occur through a group of proteins, including Raf, MEK and ERK (also known as MAPK).

The Ras-Raf-MEK-ERK/MAPK pathway helps control cell proliferation, survival, differentiation, apoptosis, motility and metabolism [23]. "Many receptors can activate Ras, which recruits, Raf kinases to the cell membrane for subsequent activation. Activated Raf kinases are the point of entry into a kinase cascade in which Raf phosphorylates and activates MEK, which phosphorylates and activates extracellular signal-regulated kinases, also known as ERK and mitogen-activated protein, or MAPK. The MAPK cascade is a three-tiered module, which is a common regulatory motif in cells. A GTPase-regulated mitogen activated kinase kinase kinase (MAPKKK) catalyzes the phosphorylation of mitogen activated kinase kinase (MAPKK), which phosphorylates MAPK, the main biological effector. Signals transmitted through protein kinase C (PKC) or Ras trigger MAPK/ERK pathway signaling, which then activate Raf1, initiating a cascade involving MEK and then MAPK/ERK activation. MAPK is a node in the network of biochemical signals. MAPK acts on other protein kinases, transcriptional factors and cytoskeletal components. These influence gene expression, metabolism, cell division, morphology and survival. So, MAPK, like many other proteins involved in signal transduction, interacts and functions in different pathways. To ensure specificity and prevent improper cross-talk, scaffold proteins can tether signaling proteins to a specific location and act as insulators, preventing inappropriate signals from reaching signaling proteins like MAPK. Some scaffold proteins undergo a conformational change when properly stimulated. In the altered conformation, signaling molecules are allowed to pass through them and reach the signaling protein, such as MAPK. So, scaffold proteins act as gates, opening the flow of information and are not just passive, static insulators" [23].

There are three major subfamilies of mitogen-activated protein kinases (MAPK): the kinases that are regulated by extracellular-signals (ERK and MAPK); c-jun N-terminal kinase (JNK); and MAPK14 [24]. "The ERK MAPK pathway is important for cell proliferation. Signals transmitted through protein kinase C (PKC) or Ras stimulate the MAPK/ERK signaling pathway, which then activates Raf1, initiating a cascade involving MEK and MAPK/ERK activation. The Ras/Raf/MEK/ERK cascade is involved in the control of growth signals, cell survival, and invasion in cancer" [20]. The stress-activated protein kinase (SAPK) or JNK is an important mediator of obesity-associated stress pathways and related metabolic deterioration [24].

Free radicals and oxidative stress can activate Ras [1, 18]. "Nitric oxide (NO) promotes the direct post-translational modification of Ras by reacting with the sulfur at cysteine 118 to form a nitrosyl group. This stimulates guanine nucleotide exchange. T-cell receptor engagement also leads to the activation of Ras through a signaling pathway involving the activation of the protein p56 and PKC (protein kinase C). Activated Ras stimulates effectors such as the Raf protein kinase family, which controls the activation of the MAPK pathway and plays a major role in controlling proliferation and differentiation. The PI3K mediates some of the Ras-dependent actin cytoskeleton remodeling and protection against apoptosis. The third Ras, RalGDS, regulates receptor endocytosis, cytoskeletal changes and DNA synthesis. The Ras-Raf-MEK-ERK pathway contains several oncogenes and is deregulated in about 30% of all human cancers. It has also emerged as a prime target for antitumor therapy" [18].

The p53 protein is a tumor suppressor [1, 25]. "It regulates the cell cycle and helps prevent cancer. Under normal conditions, there is not much p53 in the cell, but when cells are stressed, p53 becomes activated and stabilized. It can be activated by post-transcriptional modifications. When it is mutated, p53 becomes an oncoprotein and can help cause cancer. It is the most commonly found oncoprotein in human cancers. It can activate DNA repair, stop the growth of cells, and can induce apoptosis. So, when a cell is exposed to mutagens the DNA can be damaged. The normal p53 protein can either cause the DNA to be repaired, or it can cause the defective cell to die (apoptosis), preventing its growth and differentiation into cancer cells. However, p53 can cause cancer when it is mutated. So, there is a cancer vaccine undergoing clinical trials, and its target is p53" [25].

The p53 protein is also a hub in the network of cellular metabolism [1]. It interacts with many other proteins. It regulates the expression of a wide variety of genes in response to genotoxic compounds or cellular stress. The genes are involved in apoptosis, growth arrest, inhibition of cell cycle progression, differentiation and accelerated DNA repair or senescence. Activated p53 can initiate apoptosis and stop cell proliferation. When p53 is mutated and no longer active, uncontrolled cell proliferation, or cancer, can occur. Progression of a healthy cell through the cell cycle is regulated by many different proteins at several check points. Information from other proteins will affect the check point proteins. The p53 protein is tetrameric – it has four subunits. It interacts with many other biopolymers in cells. Parts of it have flexible structures, enabling it to bind to different proteins and different regions of DNA. It is a transcription factor that can activate genes to produce proteins which either halt cell cycle progression or induce apoptosis [1, 16]. "Thus, normal (or wild-type) p53 is able to sense abnormal cell function, making it an effective tumor suppressor. It can induce or suppress the expression of several genes that code for proteins involved in apoptosis, cell cycle control and senescence. It acts as a guardian of the genome when it is acting properly. However, a mutated form of p53 is the most frequently mutated oncoprotein found in human cancers. Transcription of the *TP53* gene is controlled by negative regulators called *MDM2* and *MDM4*. Even in cancer cells that have the normal, wild-type p53 gene, the activity of the p53 protein is often inhibited by MDM2, another oncoprotein. Deregultaion of the balance (homeostasis) of the ratio of p53/MDM2 can lead to the malignant transformation of cells. Small molecules, or potential drugs, are being tested for their ability to prevent MDM2-p53 interaction" [1].

The activity of the p53 protein can also be modified by phosphorylation, methylation, acetylation, glycosylation, ribosylation, summoylation, ubiquination and by interaction with other proteins, such as Mdm2 and MdmX, which are ubiquitin ligases that can also bind to the trans-activation domain of the tetrameric p53 protein [26]. "There are other proteins that can interact with Mdm2 and affect the stability and activity of p53. For many years, phosphorylation was thought to be the first step in stabilizing p53 when a cell is under stress. Phosphorylations at Ser395 and Tyr394 do inhibit Mdm2, while phosphorylations at Ser166 and Ser186 activate it. Moreover, cycles of phosphorylation-dephosphorylation and acetylation-deacetylation of Mdm2 affect its ability to inhibit p53. So, drugs that target the interactions between Mdm2 and p53 have been developed and have started clinical testing" [26].

In the second step, modified p53 binds to target genes through the conserved central, core domain, while the C-terminal region was thought to be a negative modulator that had to be modified to allow sequence-specific DNA binding [26]. "The modifications target and modify the C-terminal region of p53, thus affecting the ability of p53 to bind DNA. However, recent research indicated that both the core DNA-binding domain and the C-terminal domain of p53 possess DNA-binding activities. The core DNA-binding domain provides sequence specificity while the C-terminal domain recognizes structural features of target DNA" [26].

Acetylation is an important link between p53 and histones that regulate transcription [26]. "The same acetyl transferases, CBP/p300, Tip60 and Mof, catalyze the acetylation of p53 and histones. Mof is also a key regulator of the embryonic stem cell transcriptional network, as it primes genes for developmental programs. However, ubiquitination and acetylation are mutually exclusive. Competition between these modifications may affect the stability of p53. That is, p53 acetylation levels are markedly enhanced in response to stress, promoting p53 stabilization and activation, while ubiquination targets p53 for proteasomal degradation" [26].

Another important oncogene is *PI3K*, which codes for the enzyme PI3K (phosphoinositide 3-kinase) [1]. "It catalyzes the phosphorylation of phosphatidyl inositol.

Like other oncogenes, *PI3K* is upregulated in cancers. On the other hand, the enzyme PTEN (phosphatase and tensin homolog) catalyzes the opposite reaction, so it is a tumor suppressor and the gene coding for it is downregulated in cancer. So, PI3K and PTEN are part of a signaling system that can go wrong and lead to cancer" [1].

Not only transcription factors, but also protein receptors on the cell surface can be over-expressed. Ligands that bind to the receptors can promote cell growth and division [1, 16]. "For example, the human epidermal growth factor receptor (HER1) appears in many cancer cells. It can be either over-expressed, or it expressed in a mutated form that allows it to bind to several ligands that do not bind to the normal HER1. This receptor is a tyrosine kinase which catalyzes its own phosphorylation (auto-phosphorylation). The phosphorylated form binds to two proteins called Grb2 and Sos. These proteins associate with Ras, which activates three proteins on different signaling pathways. One of these pathways uses PI3K, which catalyzes the phosphorylation of the OH on carbon number 3 of inositol, to produce $PIP_3$. Next, $PIP_3$ can be hydrolyzed into inositol 1,4,5-trisphosphate ($IP_3$) and diacyl glycerol, which are second messengers that cause many further effects in the cell. $PI_3$ activates two other kinases, called Akt and GSK-3β, which lead to the phosphorylation of a protein called β-catenin which moves to the cell nucleus and activates gene transcription, leading to cell proliferation" [16].

Some of these proteins were mentioned when the potential health benefits of pomegranates were discussed. Another important protein is the hormone insulin. It is important in diabetes as well as breast and prostate cancer. Some hormones like estrogen are steroids. Hormones bind to receptors (another class of proteins) that can activate the transcription of genes [1]. "For example, the parathyroid glands secrete a parathyroid hormone. Fat cells (adipocytes) secrete leptin. The pancreas secretes insulin, which is the primary regulator of fat, carbohydrate and protein metabolism. Insulin regulates the synthesis of glycogen (the form in which glucose is stored in the muscles and liver) and it inhibits the synthesis of glucose in the liver. It stimulates the synthesis and storage of fats and inhibits the release of fats. Insulin also stimulates the synthesis of proteins that are needed for the proper function, repair and growth of cells. Insulin also functions as a signaling molecule. It provides information on the amount of fuel that is available to the brain and central nervous system. Insulin orchestrates the use of fuels, helping the body to decide whether to store the fuels or use them for energy. After eating a meal, blood sugar is elevated, so insulin instructs the body to store most of the fuel for later use. As insulin and sugar levels drop, insulin tells the body to start using the stored fuel for energy. When this process fails, insulin resistance develops, followed by type II diabetes, heart disease and obesity" [1].

There are also important post-transcriptional modifications (PTMs) of proteins [1]. "They help regulate protein function. Some act as on-off switches, while others mark a protein for destruction. The most common PTM is phosphorylation. Moreover, acetyls, methyls, hydroxyls, sulfate, nitrate, sugars, lipids (especially palmitoyl), and even other proteins (ubiquitin and SUMO, or small ubiquitin-like modifier) can be added to a variety of proteins" [1].

Another class of biochemicals is sugars, or saccharides [1]. "The general formula for a carbohydrate is $(CH_2O)_n$. That is, a carbohydrate has many $CH_2O$ units, or building blocks. Think of it as hydrated carbon, or carbons with waters covalently attached to them. Glucose is a simple carbohydrate. Like other sugars, it can be either the D (dextrotatory) or L (levorotatory) isomer. The D isomer of glucose is also known as dextrose. It is the only source of energy for red blood cells, and is the preferred energy source for the brain, central nervous system, placenta, and fetus. The concentration or level of glucose in the blood plasma must be carefully controlled. It should be measured after fasting >12 hrs, is called the fasting plasma glucose (FPG) test. Glucose should be below 100 mg/dL. In prediabetic people, it can be between 100-125 mg/dL. A FPG level >125 mg/dL is a diagnostic indicator for diabetes. Glucose can also be measured two hrs after giving a subject 1.75 g of glucose per kg body weight after fasting for 8-12 hrs. This is an oral glucose tolerance test and glucose should be <200 mg/dL. Levels higher than that are symptomatic of diabetes" [1].

Other common monosaccharides are fructose (in fruits), glucose, galactose, ribose (in RNA) and deoxyribose (in DNA) [1]. "Monosaccharides are the basic building blocks of complex carbohydrates. When two monosaccharides react with each other, they produce water and a disaccharide. All mono- and disaccharides are simple carbohydrates. Important disaccharides include sucrose (table sugar) and lactose. Sucrose is a disaccharide of glucose and fructose. It is the most abundant disaccharide in plants" [1].

Another sugar, called inositol, is sometimes phosphorylated in three positions, to make inositol 1,4,5-trisphosphate, or $IP_3$ [1]. "It is an intracellular second messenger in cellular communication. There are also inositol mono-, di-, tetra, penta-, hexa- and pyrophosphates (IP, $IP_2$, $IP_4$, $IP_5$, $IP_6$ and $IP_7$). Phosphates are added in reactions catalyzed by kinases and inositol polyphosphate multikinase (IPMK). $IP_6$ is also known as phytic acid, the main form in which phosphorus is stored in many plant tissues, especially bran and seeds. $IP_7$ modulates insulin sensitivity, neutrophil function, chemotaxis, endocytosis, and telomere maintenance. It also stimulates transcription mediated by the protein called p53 and enhances its acetylation. IPMK is an indispensable coactivator of p53-mediated transcription and cell death. IMPK also binds to the mammalian target of rapamycin, mTOR (also known as mechanistic target of rapamycin). It stabilizes the mTOR complex 1 (mTORC1) and increases its activity" [1]

Inositol is also part of larger molecules called phosphoinositides [1]. "They are members of the class of compounds called phospholipids, which are in cell membranes and contain a glycerol backbone, two fatty acyls and a phosphate. Inositol is hexahydroxycyclohexane. There are six asymmetric carbons, so there are $2^6$ possible isomers. The most common form is the one that occurs naturally, *myo*-inositol, or *cis*-1,2,3,5-*trans*-4,6-cyclohexanehexol" [1].

Sugars, carbohydrates, proteins, DNA, RNA and lipids all interact with water, ions and a variety of small molecules in the web of life [1, 15]. In the 20$^{th}$ century, many prescription drugs were developed in searches for molecules that could bind to a specific target, such as an enzyme or membrane-bound protein receptor. They bound to the active site of the enzyme or the binding site of the natural ligand. For example, statins inhibit an enzyme called HMG-CoA reductase, which catalyzes the rate limiting (slowest step) in the biosynthesis of cholesterol [27]. However, in recent years, statins have been found to have other important health benefits, including inhibiting angiogenesis in the lymph nodes, to where many types of cancer first spread when a tumor metastasizes [28]. More importantly, they save lives even though newer investigational drugs that inhibit cholesterol biosynthesis do not [29]. So, statins and other prescription drugs have multiple health benefits. So, it is not surprising that pomegranates do, too.

# Inflammation

"Inflammation can be a cause or a symptom of many diseases" [1]. "When controlled, it is also an important part of maintaining good health. To get energy, human cells contain mitochondria, which make pro-inflammatory free radicals and reactive oxygen substances (ROS) as by-products of the TCA cycle and oxidative phosphorylation. Other subcellular organelles, called peroxisomes, oxidize fatty acids that have more than eight carbons and produce $H_2O_2$, which is broken down by the enzyme called catalase. Before it is broken down, $H_2O_2$ can oxidize compounds such as phenols, aldehydes and alcohols. Also, it can act as a second messenger that is produced in response to extracellular stimuli, and can regulate various biological processes" [1].

Also, when disease-causing (pathogenic) microorganisms invade the body, immune cells kill the microbes by causing oxidative damage and inflammation [1]. "Although there is some collateral damage to the surrounding healthy cells and tissues, this does get repaired in healthy people. However, when the immune system is overactive, it can incorrectly identify environmental chemicals, foods and even one's own cells as foreign and mount a potentially fatal autoimmune response or allergic reaction. So, inflammation, like so much else, must be carefully controlled. Although pathogenic bacteria can cause inflammation and disease, remember that there are non-pathogenic bacteria that are essential for healthy human life. This includes *Bifidobacterium* and *Bacteriodes*, which positively influence immune responses, and protect against the development of inflammatory diseases. They do this by helping to digest fermentable dietary fiber" [1].

One definition of inflammation is that it is when parts of the body become red, warm, swollen, and damaged [1]. "This can happen when the immune system responds to foreign materials, irritation, bone or nerve damage, infection by microorganisms, and ischemia (lack of blood flow), followed by reperfusion. Inflammation is caused by the production of reactive oxygen species (or substances) and other chemicals, such as histamine, pro-inflammatory proteins (cytokines) and eicosanoids" [1].

Reactive oxygen substances (ROS) are produced as part of normal, healthy aerobic metabolism, such as electron transport in the mitochondria of cells "[1]. "Please note that much of the chemical literature uses the term reactive oxygen species (ROS), instead of reactive oxygen substances, but many biologists would consider this to be a misuse of the word species. When oxygen reacts with nutrients and metabolites, it produces ROS such as hydrogen peroxide, the superoxide anion, the hydroxyl radical and nitric oxide. Hydrogen peroxide ($H_2O_2$) is made in mitochondria as a byproduct of biochemical reactions that produce energy for cells. It can react with unchelated iron ($Fe^{2+}$) and copper ($Cu^{2+}$) to produce the superoxide anion ($O_2^-$), which can also be produced by the reaction catalyzed by NADPH oxidases in immune cells" [1].

$$2\ O_2 + NADPH \rightarrow 2\ O_2^- + NADP^+ + H^+$$

The superoxide anion can then react with other molecules to produce other ROS [1]. When a pathogenic organism enters the body, NADPH oxidase in phagocytes and T lymphocytes that originate in lymph nodes produce ROS which kill the invading organism [1].

NADPH oxidase also exists in other tissues and its activation can be harmful [1]. For example, NADPH oxidase in the lungs can cause oxidative damage after cardiopulmonary bypass operations [30]. A chemical carcinogen, diethylnitrosamine, can activate a signaling pathway in liver Kupffer cells (a type of macrophage) that causes inflammation and tissue damage [28]. In the brain, there is an NADPH oxidase in glial cells that is responsible for oxidative damage after suffering a stroke [1].

Hydrogen peroxide does not cause much direct oxidative damage, but it can react with free iron in the +2 oxidation state ($Fe^{2+}$) to produce the hydroxyl radical ($OH^{\cdot}$) in the Fenton reaction. The hydroxyl radical is highly reactive and damaging. It can oxidize lipids and lead to atherosclerosis and heart disease, as stated in the iron hypothesis [32]. That is, differences in the amounts of stored iron can explain why men are more prone to heart disease than women. Iron stores in men rise after adolescence, but remain low in women and only begins to rise after the age of about 45. The maximum sex difference in heart disease is also at age 45. Moreover, medicines like aspirin that cause gastrointestinal blood loss may protect against heart disease by decreasing iron stores. This may also partly explain the protective effect of NSAIDs on Alzheimer's disease. That is, a systems biology analysis showed that poorly liganded iron and copper (free iron and copper) can help cause not just Alzheimer's, but also Parkinson's and Huntington's disease. Free $Fe^{2+}$ and $Cu^{2+}$ can also lead to diseases caused by prions, bacteriocides and chemical toxicity [33]. That is, there is a continuous autocatalytic production of hydroxyl radicals involving poorly liganded iron, leading to cell death via apoptosis. That is, once $Fe^{3+}$ is produced in the Fenton reaction it can react with superoxide anions, $O_2^{\cdot -}$, (also produced by mitochondria) to regenerate $Fe^{2+}$ [33].

The hydroxyl radical is quite reactive and can damage many cellular components – not just lipids [33]. "It can also release $Fe^{2+}$ from Fe-S catalytic centers in proteins like ferritin, thus producing more free $Fe^{2+}$ that can react with more $H_2O_2$ to produce more hydroxyl radicals in an autocatalytic reaction. In a related reaction, the superoxide anion can react with nitric oxide (NO) to make the highly reactive peroxynitrite anion, $ONOO^-$, which can react with tyrosine or cysteine in proteins (to make nitrosylated proteins), with DNA (to make 8-hydroxy-20-deoxyguanosine) and with fatty acyls to make nitrosylated fats. So, to prevent the formation of excess hydroxyl radicals and superoxide anions, it is important to keep $Fe^{2+}$ fully liganded" [33].

That is, iron has up to six individual chelation sites, arranged octahedrally, and many ligands will bind to only some of them [33]. "For example, many dietary phenolic compounds can chelate iron and copper. However, partial chelation by ascorbate will transform vitamin C into a pro-oxidant (promoting the production of hydroxyl radicals) and not an antioxidant, which is the healthy form of it. So, the binding of incompletely liganded iron (usually bivalent) to inappropriate cellular structures can cause catalytic activities that kill cells, leading to not just atherosclerosis, but also stroke, age-related macular degeneration, prion diseases, sepsis, septic shock, viral infections and neurodegenerative diseases. It may also contribute to microbial, plant, animal and chemical toxicities" [33].

There are also some lipids, called lipoxins, resolvins and protectins that are anti-inflammatory [1]. A low dose of aspirin (81 mg) stimulates the production of lipoxins. Higher doses have no effect on lipoxin synthesis. Anti-inflammatory lipids are released in the later stages of inflammation and restore the normal, healthy condition. The successful initial inflammatory response eliminates infectious agents. This is followed by a resolution and repair phase, which is mediated by macrophages that are already in the tissue and macrophages that are drawn to the damaged site [34]. "They switch the lipid mediators from pro-inflammatory prostaglandins to anti-inflammatory lipoxins, which inhibit the recruitment of neutrophils and promote the recruitment of monocytes. They work together with resolvins, protectins, transforming growth factor β and other growth factors to remove dead cells and start remodeling tissues" [34].

There are also some endogenous inducers of inflammation that come from stressed, damaged, or malfunctioning tissues [34]. There is something called para-inflammation, or smoldering inflammation. This is a relatively low level of inflammation that occurs in diseases such as obesity, type-2 diabetes, asthma, and atherosclerosis [34]. In many cases, these diseases of inflammation can be prevented by avoiding obesity, and not consuming *trans* fats and saturated fats that are in the typical fast food diet that many people in the USA consume. Instead, unsaturated fats are much better [1].

Lipoxins are made from arachidonic acid, but the resolvins and protectins are made from omega-3 fats, such as alpha linolenic acid (ALA), eicosapentaenoic acid (EPA) and docosahexaenoic acid (DHA) [1]. "Omega-3 fats can be taken as dietary supplements such as fish oil and flaxseed oil. They are also present in fatty fish, such as salmon. Eskimos eat much fatty fish and they have a very low incidence of arthritis and heart disease" [1]. Inflammation is important in all stages of heart disease, from its beginning, development and final, tragic stages [35]. Proper diet and cardiovascular exercise can limit inflammation and help prevent heart disease.

Lipoxin A4 is important in asthma, a prevalent disease of chronic inflammation [36]. It regulates the activation of natural killer (NK) cells and type 2 ILCs (ILC2s). ILCs express high levels of cytokines that are important in the pathogenesis of the inflammatory response [36].

Inflammation plays an important role in all stages of atherosclerosis and cardiovascular disease [35]. Atherosclerosis is the main process underlying macrovascular disease. It starts when endothelial cells that line the intima are activated. This can be induced by saturated free fatty acids or cholesterol, but it leads to the expression of leukocyte adhesion molecules. So, leukocytes bind, then migrate through the endothelium to the intima where they can attract monocytes which ultimately transform into lipid-laden foam cells. This process may be enhanced in people who have type-2 diabetes. After immune cells and inflammatory mediators interact, the process can continue to atheroma and make rupture-prone atherosclerotic plaques. Inflammatory pathways are also involved in thrombosis, the late stage of atherosclerosis. It is responsible for most of the clinical complications of macrovascular disease. So, macrovascular disease can be a major consequence of obesity-induced inflammation and type-2 diabetes [37].

Smoldering inflammation can also lead to diabetes, stroke and cancer, along with neurodegenerative and autoimmune diseases. Inflammation may be more important than cholesterol levels in causing heart disease. Fifty percent of all heart attacks and strokes occur in people with normal or even low levels of cholesterol [38]. Normal, healthy endothelial cells (ECs) on the innermost surface of arterial walls resist adhesion by leukocytes. When a person smokes, consumes many saturated fats, is obese, has high blood pressure, is hyperglycemic or resistant to insulin, adhesion molecules (such as vascular cell adhesion molecule-1, or VCAM-1) are expressed by ECs. This allows leukocytes to attach to the arterial wall. VCAM-1 binds to monocytes and T lymphocytes, which are found in early atherosclerotic plaques. Oxidized lipids, nuclear factor-κB (NF-κB) and TNF-α can also stimulate VCAM-1 expression. These things can be prevented by laminar blood flow, which causes some anti-atherosclerotic mechanisms, such as expression of the natural anti-oxidant, superoxide dismutase and an increase in nitric oxide (NO) synthetase. There is an increase in the concentration of NO, which causes vasodilation and limits VCAM-1 gene expression [35].

However, when blood flow is disturbed, lesions can form and monocytes can adhere to the arterial endothelium [35]. "Monocytes can then enter the vessel walls, causing a fatty streak to start developing. This is the first stage of atherosclerosis. Monocytes mature into macrophages, which express scavenger receptors and engulf modified lipoproteins. Cholesteryl esters accumulate in the cytoplasm and macrophages become foam cells, which have an increased concentration of lipids. Next, T-lymphocytes of the adaptive immune system enter the arterial cell walls and participate in further inflammation. They are attracted to the arterial cells by chemokines (γ-IP-10, MIG and I-TAC) that are induced by interferon-γ" [35].

The next stage is progression to the formation of plaque, which can begin early in life, when a person is still in their early twenties [14]. A mixture of cytokines: interleukins (IL-1β, IL-1α, IL-6, IL-18) and tumor necrosis factors (TNF-α, TNF-β) are involved in this process, as are the macrophage colony stimulating factor (MCSF) and the monocyte chemoattractant protein, MCP-1. Next, the plaques can rupture, leading to thrombosis. In healthy arteries, there is a fibrous cap that contains collagen for strength. The strong cap protects blood from contacting the lipid core. When the cap ruptures, a thrombus, or blood clot, can form. Inflammation interferes with the integrity of the collagen matrix by blocking the creation of new collagen fibers and by stimulating the destruction of existing collagen. T-lymphocytes participate in this pro-inflammatory process. They produce CD40, a cell surface protein and IL-1. These biomolecules promote the production of enzymes in macrophages that catalyze the destruction of collagen. The CD40 ligand also stimulates macrophage production of tissue factor VII, which initiates the coagulation cascade.

To prevent all these things from happening, one should try to lower the risk factors by not smoking, engaging in regular cardiovascular exercise and by consuming a healthy diet, with fewer calories, and more fresh fruits and vegetables and other sources of antioxidants [35]. An important biomarker for generalized inflammation is the high-sensitivity C-reactive protein, or hsCRP, which binds to phosphatidyl choline that is present on the exterior surface of dead or dying cells and activates the complement system of the innate immune system [35]. In healthy postmenopausal women, it is a more accurate predictor of risk for a heart attack than total cholesterol or LDL [38].

Obesity is also linked to chronic, low-grade inflammation in adipose and liver tissues [39]. "This can lead to insulin resistance and type-2 diabetes through the pro-inflammatory nuclear factor-κB (NF-κB), which is activated by the phosphorylation of the regulatory protein IκB by IκB kinases (IKKs). One of these kinases, IKKε, is elevated in adipose tissue when on a high fat diet. In an attempt to identify a small molecule inhibitor of IKKε, a library of 150,000 compounds was screened and one of them (amlexanox) bound with high affinity to the IKKε receptor. This drug is currently used to treat asthma allergic rhinitis and aphthous ulcers. When given to mice on a high fat diet, it prevented weight gain, improved insulin sensitivity, attenuated hepatic steatosis, reduced adipose tissue inflammation and promoted energy expenditure in adipose tissue through increased thermogenesis" [39].

Inflammation is also important in neurodegenerative diseases. Omega-3 fats are very important anti-inflammatory nutrients that are important in the brain, where they play an important role in cognitive function and behavior [1]. The typical American (USA) fast food diet contains little or no omega-3 fats [40]. Instead it contains high levels of *trans* fats and saturated fats, which increase the risk of heart disease and may be partly responsible for the poorer academic performance of children in the USA, compared to other countries that don't eat so much unhealthy food [40].

Inflammation is also an important factor in Alzheimer's disease [16, 41, 42]. This was shown recently in a continuation of the Framingham study [41]. "The goal of the Framingham study was (and still is) to monitor the health of thousands of people for many years, and for several generations. Even though the original goal was to see if there were certain lifestyles that made a person more or less likely to suffer from cardiovascular disease, it now looks at Alzheimer's disease and dementia as well. Scientists and doctors have monitored the health of thousands of people for many years, and for several generations. They found that people who had higher concentrations of pro-inflammatory cytokines in their blood were more likely to get Alzheimer's disease" [41]. It has also been shown that people who take NSAIDs earlier in life have a lower incidence of Alzheimer's disease [42]. However, once a person gets Alzheimer's disease, NSAIDs are no longer effective.

Inflammation is also an important factor in stroke, which can be caused by the temporary lack of blood flow to the brain (ischemia), followed by re-perfusion. Ischemia-reperfusion causes glia and astrocytes to become activated and produce ROS, which cause oxidative damage and inflammation [1]. Antioxidants can prevent this from happening [43].

Reactive oxygen substances (ROS) and inflammation are probably involved in cancer, too [1, 16]. In the 19$^{th}$ century, doctors observed that tumors often arose at sites of chronic inflammation [16]. In the 20$^{th}$ and 21$^{st}$ centuries, epidemiological studies showed that chronic inflammation is an important risk factor in many types of cancer [16]. At the same time NSAIDs have been shown to help prevent cancer [16]. Damage to DNA by ROS is a widely accepted cause of cancer [44]. "Most mutations caused by ROS convert deoxyguanosine into deoxythymidine (G $\rightarrow$ T). Another mutation is the accumulation of 8-hydroxy-deoxyguanosine, or 8-OHdG, which can be caused by nitrosamines and polynuclear aromatic hydrocarbons produced by smoking tobacco. Some viruses, such as hepatitis B and C, can cause chronic inflammation and liver cancer. Also, G $\rightarrow$ T transitions can be caused by aflatoxins, which are known human carcinogens. Some types of cancer cells also produce ROS of their own. ROS production can be induced by oncogenes, such as *Ras*" [44]. Also, pro-inflammatory cytokines, such as IL-1β, IL-6, IL-23 and TNF-α are produced after genes are activated by transcription factors, such as transcription factor nuclear factor NF-κB and STAT3, or signal transducer and activator of transcription activator 3 [16]. Phase I and II trials of antagonists of IL-6 and other chemokines are being conducted [44]. NF-κB is a coordinator of innate immunity and inflammation and is a tumor promoter. It activates the transcription of genes that code for pro-inflammatory cytokines, adhesion molecules, enzymes (such as COX and inducible nitric oxide synthase, iNOS) and angiogenic factors [16].

There are tumor-associated macrophages, or TAMs, which infiltrate most, if not all, tumors [16]. They usually have an M2 phenotype, which promotes tumor growth, remodels tissues, promotes angiogenesis and suppresses adaptive immunity [16]. Angiogenesis is required to supply blood to tumors after they grow to a certain size. Vascular endothelial growth factor (VEGF) not only stimulates angiogenesis, but also attracts monocytes into the primary tumor and the metastatic tumor. Then the monocytes secrete more pro-angiogenic factors, further stimulating angiogenesis. Chemokine receptors and their ligands also direct the movement of tumor cells during inflammation. When normal cells are transformed into cancer cells, they start to express chemokine receptors that help cause cell migration and survival at their metastatic site" [16].

So, inflammation can be good or bad [1]. It is needed to kill invading bacteria, as part of a healthy immune response. It is bad when it is unresolved, in what is often called smoldering inflammation. This occurs during obesity and chronic infections and can be caused by periodontal disease. Omega-3 fats and polyunsaturated fats help protect against smoldering inflammation because they are converted into resolvins and protectins, which resolve the smoldering inflammation and protect against further inflammation. So, omega-3 fats and polyunsaturated fats can help prevent diseases of inflammation, including arthritis, cancer, diabetes, heart disease, stroke, Alzheimer's disease, hormonal diseases, osteoporosis, inflammatory bowel disease, pelvic inflammatory disease, and many other diseases´[1].

Inflammation is a significant factor in many diseases, including arthritis, cancer, diabetes, heart disease, stroke, Alzheimer's disease, hormonal diseases, osteoporosis, inflammatory bowel disease, pelvic inflammatory disease, and many others. Smoldering inflammation is a relatively low level of inflammation that occurs in obesity, type-2 diabetes, asthma, and atherosclerosis [1]. "In many cases, these diseases of inflammation can be prevented by avoiding obesity, *trans* fats and saturated fats that are in the typical fast food diet that many people in the USA consume. Instead, unsaturated fats are much better. Omega-3 fats are also very important in the brain, where they play an important role in cognitive function and behavior. Inflammation plays an important role in all stages of atherosclerosis and cardiovascular disease. Inflammation is also an important factor in stroke" [1].

## IMMUNE SYSTEM

The immune system recognizes and defends us against internal threats caused by invading organisms and pathogens [1]. "The innate immune system recognizes bacteria, fungi and other organisms, breaks them down, identifies a characteristic protein on them (an antigen) and attaches it to the surface of specific cells, which present it to the adaptive immune system for destruction. The adaptive or acquired immune system acts once it is stimulated by the innate system. The adaptive immune response is initiated by specific interactions between antigen-loaded, mature dendritic cells and naïve $CD4^+$ T cells in the lymph nodes. The adaptive or acquired immune system acts once it is stimulated by the innate system. The five families of immune cells are: phagocytes, granulocytes, natural killer (NK) cells, lymphocytic T-cells and lymphocytic B-cells. The four major classes of immune system mediators are chemotactic agents, cytokines, C-reactive protein and antibodies. A fifth class of immune network mediators are the small molecules, including neurotransmitters, such as L-DOPA and catecholamines" [1].

It is essential that we can distinguish between self and non-self [1]. "This is important because we are constantly being exposed to many human-made chemicals, viruses and organisms that are potentially very dangerous. From the outside of our skin to the inside of our bodies, we have many ways to defend ourselves. Similarly, all organisms have defenses. Both invertebrates and vertebrates have an innate immune system, but only vertebrates have an acquired (or adaptive) immune system. In fact, it is often described as an immune network" [1].

Our first line of defense is composed of physical and chemical barriers, such as antimicrobial peptides and enzymes (like lysozyme) that are secreted by saliva, tears, the respiratory tract and the skin (a physical barrier) [1]. "In the gut, commensurate (or commensal) bacteria help prevent infection by pathogenic bacteria. The ecosystem that comprises the human organism co-evolved with commensurate bacteria that have established a niche for themselves, in which they have a competitive advantage over other bacteria. The highest density of commensurate bacteria in the human body is found in the gastrointestinal tract" [1]. There is a mucus tract that covers it and influences the function of antigen presenting cells (APCs) and epithelial cells so that our dendritic cells (DCs) can tolerate food and commensurate bacterial antigens [45]. This mucus tract contains glycoproteins called mucins, including MUC2. It prevents inflammation by forming a non-attached outer mucus layer that is inhabited by bacteria. MUC2 also forms an inner mucus layer that adheres to intestinal epithelial cells. It mitigates inflammatory responses to DCs by generating signals that make them tolerant to antigens [46]. DCs lie just beneath the epithelial layer of cells and presents foreign antigens which we should not try to tolerate to other cells in the immune system [1].

When these first lines of defense fail to block pathogenic bacteria or other immunogens, the innate and adaptive immune systems are activated [1]. "Both use white blood cells (leukocytes), which are a diverse group of cell types that mediate the body's immune response. They circulate through the blood and lymphatic system and are recruited to sites of tissue damage and infection" [1].

Acute inflammation is made up of antibodies and humoral factors in the cell-free blood serum and other bodily fluids, once known as humors [1]. This includes the complement system, coagulation system, iron-binding proteins (lactoferrin and transferrin), interferons, lysozyme and a protein called interleukin-1, or IL-1. Antibodies are made in B cells, which are lymphocytes that are made in the bone marrow. The innate immune system recognizes pathogen-associated molecular patterns (PAMPs), which include double-stranded RNA, lipopolysachharides and other molecules that are common in viruses, bacteria and fungi, but not in healthy human cells. The innate immune system senses the presence of microbial pathogens by detecting cytoplasmic DNA. However, erroneous detection of dinucleotides can help cause some autoimmune diseases [1].

The innate immune system also recognizes damage- or danger-associated molecular patterns (DAMPs) released from damaged cells, such as uric acid crystals or ATP [1]. "It protects the host from infection by other organisms in a relatively non-specific way. The human innate immune system recruits immune cells to the site of infection, activates the complement cascade, identifies and removes foreign substances and activates the adaptive immune system through antigen presentation. That is, the innate immune system recognizes bacteria, fungi and other organisms, breaks them down, identifies a characteristic protein on them (an antigen) and attaches it to the surface of specific cells, which present it to the adaptive immune system for destruction. The innate immune system does not have a memory and does not offer specific resistance against organisms that have invaded the host in the past.

For this, the adaptive immune system is needed. It is activated by the innate immune system. It is made of many parts" [1].

Invading cells are killed by phagocytosis using a subgroup of leukocytes that circulate in the blood as monocytes [1]. "They are converted into macrophages, which enter tissues during inflammation. There is much heterogeneity in phenotype, homeostatic turnover and function when they are present in different tissues. Dendritic cells (DCs) are a distinct lineage of mononuclear phagocytes, specialized in presenting antigens to T cells while initiating and controlling immunity. DCs are crucial for activating and conditioning virus-specific T cells" [1]. They bridge innate and adaptive immunity and have additional roles in shaping the immune response to pathogens, vaccines, and tumors [47].

Phagocytic white blood cells have several cell surface protein receptors [1]. "This include $F_c$ receptors that bind to the $F_c$ (crystallizable fraction) of IgG antibodies; complement receptors, which bind complement proteins; scavenger receptors, which bind a variety of polyanions on bacterial surfaces; and toll-like receptors, which recognize PAMPs on infectious agents" [1].

There is also a cell-based immunity that does not include complement or B cells [1]. "Instead, cell-mediated immunity works through macrophages, natural killer (NK) cells, cytotoxic T-lymphocytes (made in the thymus) and many different cytokines that are released in response to an antigen, bacteria, fungi, protozoans and cancer cells. Both the innate and adaptive immune systems use cell-mediated and humoral components" [1].

The adaptive or acquired immune system acts once it is stimulated by the innate system [1]. "Mature dendritic cells in the innate immune system form immunological synapses with T cells that have the CD4 cell surface glycoprotein, or $CD4^+$ T-cells. There are two main types of T cells: $CD4^+$, which oversee the immune response and $CD8^+$ cells which do most of the actual killing. The adaptive immune response is initiated by specific interactions between antigen-loaded, mature dendritic cells and naïve $CD4^+$ T cells in the lymph nodes. Dendritic cells link the innate and adaptive immune systems. They engulf exogenous pathogens and toxins, chop them into pieces and present them to T-cells, which come from the thymus gland. To do this properly, they must be able to distinguish between self and non-self. Although our brains may establish our personal identities to the outside world, the immune system establishes our internal identities" [1].

As the name implies, the adaptive immune system adapts to toxic challenges from viruses, bacteria, protozoa, and other organisms [1]. "Adaptive immunity uses helper T cells ($T_H$) that produce cytokines. Naïve $T_H$ cells are stimulated by antigen presenting cells that have cognate antigens on their surface. The naïve cells differentiate into two different types of $T_H$ cells, $T_H1$ and $T_H2$. The $T_H1$ cells secrete interferon-γ and promote cellular immunity. The $CD4^+$ $T_H2$ cells produce the interleukin (IL) cytokines IL-4, IL-5, IL-10 and IL-13, while producing humoral immunity. A small percentage of the cells acquire and retain a memory of one or more pieces (antigens) of the invading viruses and micro-organisms. This enables us to mount rapid and effective responses to the invading pathogens when they invade us more than once" [1].

The adaptive immune system uses the type of white blood cells that are called lymphocytes [1]. "The human body has about 2 trillion lymphocytes, constituting 20-40% of all the white blood cells. The peripheral blood contains 20–50% of them as circulating lymphocytes; the rest move within the lymphatic system. Natural killer cells are large and granular. Smaller lymphocytes are T and B cells. There are three general classifications of

$CD8^+$ cytotoxic T cells: naïve cells, central memory and effector memory cells. Naïve T-cells have not yet seen their cognate antigen, but memory cells have. Memory T cells can recognize bacteria, viruses and cancer cells. Central memory cells express the cytokine CCR7 and interleukin IL-2. Effector memory T cells ($T_{EM}$) express the effector cytokines IFNγ and IL-4. The $T_{EM}$ cells can reside permanently (or with low turnover rates) in peripheral tissues and migratory cells that can move from the periphery and into the blood. Moreover, $T_{EM}$ cells can rapidly recognize and kill infected target cells. They can facilitate early containment of nascent infections" [1].

## METABOLIC SYNDROME, DIABETES, HEART DISEASE AND STROKE

Heart disease is the biggest killer in the world. Its major cause is obesity or metabolic syndrome, which can also lead to diabetes and stroke - the third leading killer (after cancer) and the most frequent cause of disability in the world [1]. "All of these involve imbalances in energy metabolism. Early symptoms of metabolic syndrome include excessive weight, high blood pressure, elevated blood glucose and elevated levels of lipids, especially low density lipoprotein (LDL). People who have two or more of these symptoms are at a higher risk of developing heart disease, stroke and type-2 diabetes. Childhood obesity is such a big problem that it will probably mean that this generation will be the first to have a shorter life expectancy than their parents. Ischemic strokes need to be treated as soon as possible, so it is very important to get a stroke victim to a hospital as soon as possible, so the blockage can be removed and blood can begin to flow properly. Drugs such as aspirin, clopidrogel and dipyridamole can be given to prevent more clots from forming by preventing blood platelets from aggregating. If the patient can get to a hospital soon enough, tPA, or tissue plasminogen activator, can be given. When blood flow to the heart is interrupted, it causes a myocardial infarction, or heart attack. The most common cause is a blockage in the coronary artery that is usually caused by the rupture of an atherosclerotic plaque" [1].

Obesity and metabolic syndrome can lead to diabetes and stroke, which is the third leading killer (after cancer) and the most frequent cause of disability in the world [1].

For a person to be healthy, it is important that she (he) is able to convert food into needed energy [1]. "Dietary carbohydrates are converted to glucose, which is the primary fuel that provides energy to our cells (especially in the brain), and as a source of intermediates for the biosynthesis of other carbohydrates, as well as nucleic acids, fats, amino acids and proteins. Glucose homeostasis is maintained by several hormones, with insulin and glucagon being the most important. When the concentration of glucose in the blood rises in a healthy person, the β-cells of the pancreas secrete insulin, which inhibits hepatic synthesis of glucose and/or increases glucose uptake into the muscles, liver and adipose tissue. This causes blood glucose levels to decrease. Glucagon is secreted by the α-cells of the pancreas when the concentration of glucose in the blood is low. Glucagon has the opposite effect of insulin. It stimulates the liver to make more glucose by breaking down some of its stored glycogen" [1].

Glucose can be broken down further by glycolysis to produce energy, or it can be converted into glucose-3-phopshate and then into phospholipids and/or triglycerides, depending on the needs of the cell [1]. "Triglycerides can be stored for later use, or hydrolyzed into glycerol and free fatty acids, which can be used for fuel. They are broken down to acetyl-CoA, and fed into the citric acid cycle. When energy is no longer needed, fatty acids can react with glycerol to form triglycerides, which store energy" [1].

The mitochondria are often called the power plants of the cell [1]. "That is, one of their major functions is to produce energy in the form of adenosine triphosphate, or ATP. They do this in an electron transport chain. NADH and $FADH_2$ are oxidized and ATP is produced, along with reactive oxygen substances (ROS). ATP activates many biomolecules and powers much of the biosynthesis, development, regulation and repair that are required to sustain life. As a result, inefficiency or poor regulation of energy metabolism can have a profound effect on human health. However, natural history and human history can work against us. We evolved with an ability to store energy at times when food was not plentiful. So, when modern humans go on a diet, our genes code for proteins which slow down our metabolism and limit our ability to lose weight" [1].

Metabolic syndrome was first described by G.M. Reaven in 1988 [1]. In addition to the symptoms just mentioned, metabolic syndrome is characterized by many abnormalities, including inefficient use of glucose as an energy source, excessive conversion of glucose to fat, lowered sensitivity to insulin, loss of mitochondrial function and energy production, low levels of high density lipoprotein (HDL), excessive appetite and weight gain, impaired circulation because of clogged arteries, inflammation, impaired organ function and depressed immune response [1].

Adipose tissue is an active endocrine and immune organ. It secretes lipid hormones (adipokines). When caloric intake exceeds energy consumed, adipocytes undergo increased recruitment, proliferation and differentiation. If adipogenesis is sufficient, metabolic syndrome can be avoided. When adipogenesis is impaired, adipocytes can become dysfunctional and help cause diseases related to metabolic syndrome. This is often called adiposopathy, or sick fat. This fat is especially dangerous when it accumulates in the abdomen. Adiposity increases the macrophage content of adipocytes.

Metabolic syndrome and smoldering inflammation can also lead to type-2 diabetes, especially when a person has excess visceral, abdominal deep subcutaneous and ectopic fat [37]. The anti-inflammatory hormone insulin plays an important role in the development of type-2 diabetes [1]. When insulin binds to its receptor, it initiates a signaling cascade that must function properly to maintain good health. On the other hand, the $\alpha_{2A}$ adrenergic receptor mediates the suppression of insulin secretion, when it is appropriate [48]. When the signaling cascade is not regulated properly, it can contribute to metabolic syndrome, even before the onset of diabetes [1]. Insulin resistance and a decreased capacity for $\beta$ cells to secrete insulin are central features of type-2 diabetes.

It was estimated that 346 million people world-wide have diabetes [12]. It is characterized by hyperglycemia (fasting glucose >125 mg/dL) and an elevated percentage of glycated hemoglobin ($\geq$6.5%), known as $HbA_{1c}$. That is, a chronic excess of glucose can cause tissue damage. Some of it is converted into fructose 6-phosphate, which feeds into the hexosamine pathway, which produces N-acetylglucosamine that can react with serine and tyrosine residues in proteins that are important in signal transduction. This is called OGlcNAcylation. Also, some of the excess glucose is converted to another sugar, sorbitol, as it enters the polyol pathway in a reaction catalyzed by aldol reductase, as shown in the following reaction.

Glucose + NADPH + $H^+$ → Sorbitol + $NADP^+$

In the process, NADPH is converted to $NADP^+$, reducing the amount of NADPH that is needed to synthesize reduced glutathione, nitric oxide, *myo*-inositol and taurine. Nitric oxide (NO) is a vasodilator in blood vessels. In healthy endothelial cells, insulin activates endothelial nitric oxide synthase (eNOS). In diabetes, endothelial insulin resistance impairs eNOS activation, reduces endothelium-dependent vasodilation, and promotes atherogenesis [13]. Moreover, when the endothelium is inflamed due to activation by the nuclear factor kappa B (NKκB), insulin-mediated signaling pathways are perturbed, reducing NO bioactivity and increasing the release of cytokines [13]. Also, glutathione is needed to destroy reactive oxygen and nitrogen substances (RONS, more commonly called reactive oxygen species, ROS) that are made by mitochondria as they produce energy. The elevated levels of RONS (or ROS) cause oxidative damage and smoldering inflammation. They can react with nitric oxide (NO) to make an especially reactive substance, peroxynitrate ($ONOO^-$), which can react with proteins, nitrosylating them. Glucose can be oxidized to methylglyoxal, which can cause proteins (like hemoglobin) to be glycated. These are called advanced glycation end products, or AGEs. Lipid oxidation can also form AGEs, including the most potent one, methylgyoxal. Another AGE, N-(Carboxymethyl)lysine (CML) accumulates in adipose tissue and fatty liver and provided evidence that this is a core mechanism leading to the dysregulation of cytokines production. CML is a major ligand for the receptor for AGE (RAGE). In addition to the effects in insulin resistance, AGEs have also been shown to induce β-cell dysfunction and apoptosis. AGEs can induce lipid peroxidization, damage mitochondria and cause apoptosis in cardiomyocytes. Sorbitol can also produce AGEs which can alter protein function and bind to specific receptors, changing of gene expression and causing toxicity. AGEs can also cause diabetic cardiomyopathy by activating the βII isoform of protein kinase C (PKC). This, in turn, alters cardiomyocyte contraction and the amount of $Ca^{2+}$ that can be released from the endoplasmic reticulum [14]. Furthermore, AGEs stimulate the formation of more RONS (ROS), such as the superoxide anion. Moreover, sorbitol cannot cross the cell membrane, so it builds up in the cell. This causes osmotic stresses as water enters the damaged cells. Sorbitol enters the sorbitol-aldolase reductase pathway, also known as the polyol pathway. When excess glucose is converted to sorbitol, less inositol is made. It is needed to make inositol phosphates (like $IP_3$) and phosphoinositides, which are especially important for properly functioning nerves and brain cells. Also, nitrosylated proteins and oxidized lipids are toxic and linked to neurodegenerative diseases and stroke.

Finally, excess glucose can lead to an increase in diacylglycerol (DAG), which activates PKC, which then stimulates the production of RONS. That is, glucose can enter glycolysis and produce an intermediate, dihydroxyacetone phosphate, DHAP. This can be converted to glycerol 3-phosphate, which can react with two fatty acids to form diacyl glycerol 3-phosphate. After the phosphate is hydrolyzed off, DAG is formed. In the vasculature, DAG and RONS (ROS) activate the β isoform of PKC. This leads to changes in cell signaling, the production of vasoconstrictors, and conversion of smooth muscle and endothelial cells to a proliferative phenotype in the retinal microcirculation and peripheral conduit vasculatures [15]. The activated DAG-PKC pathway can lead to vascular abnormalities in the retinal, renal, neural and cardiovascular tissues [16]. This pathway regulates endothelial permeability, vasoconstriction, extracellular matrix (ECM) synthesis and turnover, cell growth, angiogenesis, cytokine activation and leucocyte adhesion in the heart. Increased DAG levels and PKC activity occur in diabetes. This is especially true for the α, β1/2 and δ isoforms of PKC in the retina, aorta, heart, renal glomeruli and circulating macrophages.

Activated PKC has been associated with changes in blood flow, basement membrane thickening, extracellular matrix expansion, increases in vascular permeability, abnormal angiogenesis, excessive apoptosis and changes in enzymatic activity alterations such as $Na^+/K^+$–ATPase, cytosolic phospholipase 2 (cPLA2), PI3 Kinase and mitogen activated protein kinase (MAP kinase). Inhibition of PKC, especially the β-1/2 isoform, has been reported to prevent or normalize many vascular abnormalities [16]. Clinical studies have shown that ruboxistaurin, an orally available selective inhibitor of the PKCβ isoform, can normalize endothelial dysfunction, renal glomerular filtration rate and prevented loss of visual acuity in diabetic patients [17]. However, the FDA mandated another three-year clinical trial on ruboxistaurin, which the manufacturer may be reluctant to do, because of the cost. Still, they are looking into other indications for ruboxistaurin. Two phase III studies and one phase II study are ongoing in the USA. Two of these are looking at the effects of ruboxistaurin treatment on diabetic macular edema in patients with Type-1 and Type-2 diabetes. The other is measuring the effects of ruboxistaurin treatment on early diabetic kidney in patients with Type-1 diabetes [18]. Eventually, diabetes can lead to heart disease, stroke, loss of vision, retinopathy, kidney failure, nervous system damage and even death [19].

Type-1 diabetes, also known as juvenile onset diabetes and insulin dependent diabetes, is the most severe form [1]. "It is an autoimmune disease in which the body's immune system destroys the beta (β) cells in the pancreas. That is, the pancreas has endocrine and exocrine tissues. The exocrine cells secrete digestive enzymes which are delivered to the intestines. Clusters of endocrine cells (called islets of Langerhans) exist in the exocrine tissue. Within the islets, α-cells make glucagon; β-cells, insulin; δ-cells, somatostatin; and γ-cells, pancreatic polypeptide — all of which are delivered into the blood stream. Insulin and glucagon have opposite effects. Glucagon raises blood glucose and causes the liver to convert glycogen to glucose, which is released into the bloodstream. So, after eating a meal, when blood glucose increases, β-cells secrete insulin, which lowers the glucose. If levels fall too much, the α-cells secrete glucagon, elevating the glucose level. Obesity is correlated with insulin resistance, leading to type-2 diabetes. However, type-1 diabetes can affect infants and children long before they are old enough to become obese" [1].

Smoldering inflammation is characterized by elevated levels of circulating proinflammatory cytokines and free fatty acids [1]. "They can help to induce insulin resistance, and β-cell dysfunction. The sensitivity of target cells to insulin decreases. TNF-α and lipopolysaccharides can induce insulin resistance and the level of signal transduction through a physiological negative feedback mechanism of normal insulin signaling" [1]. When insulin binds to its receptor, tyrosine residues in the intracellular part are autophosphorylated. In the metabolic pathway of insulin signaling, the insulin receptor substrate (IRS) docks with the insulin receptor and is trans-phosphorylated in its tyrosine residues by the phosphorylated insulin receptor. Subsequently, more members of the insulin signal transduction pathway, including phosphatidylinositol-3-kinase (PI3K) and Akt/protein kinase B (PKB), are recruited and activated, so they can induce downstream effects [37].

The physiological negative feedback mechanism is induced when insulin activates mTOR and the ζ isoform of protein kinase C, PKCζ [37]. "These intracellular serine-threonine kinases can then either directly, or indirectly (e.g. via IkappaB kinase beta (IKKβ)), phosphorylate serine and threonine residues in the insulin receptor substrate, or IRS. Once it is phosphorylated, IRS interrupts, or at least reduces, insulin signal transduction, and induce dissociation of IRS proteins from the insulin, receptor, induce degradation of IRS proteins, remove IRS proteins from complexes that keep them in close proximity to the insulin receptor, and turn IRS proteins into inhibitors of insulin receptor kinases. The proinflammatory cytokines TNF-α, IL-670 and IL-1α, and saturated free fatty acids are all involved in obesity-associated low-grade inflammation. They can induce phosphorylation of the IRS proteins and disrupt insulin signaling, but with pathophysiological consequences. Obese, hypertrophic and/or insulin resistant adipocytes release saturated fatty acids that can activate the TLR-4/NF-kB pathways on macrophages in adipose tissue, which then release TNF-α, which in turn binds to TNF receptors on the adipocytes. This further stimulates fatty acid release and starts a vicious cycle of worsening inflammation and insulin resistance. However, unsaturated and omega-3 fats are anti-inflammatory" [37].

Insulin signal transduction by PI3 kinase mainly affects metabolic pathways such as translocating the glucose transporter (GLUT4 in the heart) and inhibiting hormone-sensitive lipase [37]. "The other main insulin signal transduction pathway of is the renin-angiotensin system/mitogen activated protein (RAS/MAP) kinase pathway. It primarily stimulates mitogenic rather than metabolic processes. The renin-angiotensin system (RAS) is a hormone system that regulates blood pressure and fluid (water) balance. When the blood volume is low, the kidneys activate prorenin and secrete renin into the blood plasma, where it catalyzes the conversion of angiotensinogen (released from the liver) to angiotensin I. It is then converted to angiotensin II in a reaction catalyzed by the angiotensin converting enzyme (ACE). It is vasoactive and causes blood vessels to constrict, raising the blood pressure. Moreover, angiotensin II stimulates the adrenal cortex to secret aldosterone and the RAS/MAP pathway in the vasculature. This causes the tubules of the kidneys to increase the reabsorption of $Na^+$ and water into the blood. This increases the volume of fluid in the body, and thus the blood pressure. If the RAS is too active, it can cause hypertension (high blood pressure). There are many anti-hypertensive drugs that target the RAS. They are also prescribed to treat heart failure, kidney failure and to the harmful effects of diabetes" [37]. Please note that RAS is quite distinct from the oncoprotein ras, or the gene that codes for, *RAS*.

Circulating proinflammatory cytokines and free fatty acids can also cause macrophages in adipose tissue to develop into pro-inflammatory macrophages, called $M_1$ macrophages [37]. "However, there are also anti-inflammatory $M_2$ macrophages that resolve or dampen inflammation caused by $M_1$ macrophages. In obesity, most of the macrophages in adipose tissue are the $M_1$ phenotype. Then, as triglycerides build up in the liver, steatosis (fat accumulation) can begin and lead to non-alcoholic fatty liver disease (NAFLD). This can develop further into steatohepatitis, fibrosis and eventually cirrhosis of the liver. Excess visceral fat exposes the liver to higher concentrations of proinflammatory cytokines and free fatty acids. They are transported through the portal vein and into the liver, where they contribute to the development of NAFLD. So, the occurrence and severity of visceral fat accumulation are highly correlated with NAFLD. Also, excess visceral fat may contribute directly to smoldering inflammation and increased systemic levels of free fatty acids. On the other hand, there is a protein hormone, adiponectin, which has anti-inflammatory effects in adipose tissue. It may exert these effects by regulating microRNAs that suppress intracellular proinflammatory pathways, such as toll-like receptor 4 signaling. So, microRNAs have become a potential therapeutic target or tool to treat insulin resistance and type-2 diabetes" [37].

Metabolic syndrome also important causes changes in cells and tissues [37]. "Healthy β-cells can maintain glucose homeostasis by increasing their size. However, when demand for insulin exceeds the ability to produce it, hyperglycemia can occur. It can start with β-cell failure due to acquired and/or genetic defects in existing β-cells or by reduced β-cell mass. Prolonged exposure of pancreatic β-cells to toxic levels of glucose and lipids (glucotoxicity and lipotoxicity) can cause oxidative stress. Inflammatory cytokines may also contribute to β-cell dysfunction and type-2 diabetes. They cause β-cells to produce IL-1β, which can cause the β-cells to deteriorate. Also, leptin, which circulates in considerably increased concentrations in obesity, was shown to increase the release of IL-1β by β-cells" [37].

At the same time, microvascular dysfunction is involved in a cycle of both causing type-2 diabetes and being caused by it [37]. "Obese insulin-resistant humans and rats have impaired capillary recruitment, which has been shown to be necessary for normal insulin-mediated glucose uptake by skeletal muscle. Elevated levels of free fatty acids and inflammatory cytokines, together with less adiponectin can lead to dysfunctional microvasculature. This can induce endothelial insulin resistance, reduce local NO production and lower insulin-mediated glucose uptake in muscle, as part of overall insulin resistance. Endothelial dysfunction: a shared factor underlying both micro- and macrovascular dysfunction. Macrovascular disease is a major consequence of obesity-induced inflammation and type-2 diabetes. Microvascular dysfunction may be further aggravated in the expanding adipose tissue, since it produces all the factors of the renin- angiotensin system (RAS) that are needed to produce angiotensin II. Moreover, RAS activity is enhanced in obesity. Perivascular fat around resistance arterioles of muscle may directly affect the function of these vessels. Microvascular dysfunction may also contribute to the vicious cycle of adipose tissue dysfunction and inflammation. Functional capillaries in the expanding adipose tissue are necessary to provide optimal blood flow as well as deliver nutrients and oxygen to adipocytes. Thus, insufficient adipose tissue angiogenesis and formation of capillaries may lead to hypoxia and induction of an inflammatory response. A relative reduction in the density of the capillary network, combined with microvascular dysfunction may aggravate the hypoxic and inflammatory processes in adipose tissue depots and lead to deterioration of insulin resistance and metabolic homeostasis. Microvascular dysfunction may also contribute to the development of type-2 diabetes due to its effects on β-cell function. For example, there are transient periods of (mild) hyperglycemia that coincide with insulin resistance and low-grade inflammation. This can lead to reduced perfusion and mild ischemia in the islets" [37].

Moreover, several processes induce the inflammatory response in adipose tissue [37]. "This includes adipocyte hypertrophy, apoptosis and macrophage infiltration, which most likely act simultaneously. When activated by the appropriate death ligands, receptors induce apoptosis that is mediated by caspase-8. If caspase-8 activity is blocked, however, then these receptors promote death by necroptosis (programmed necrosis). Necroptosis promotes inflammation, so inhibiting this pathway can limit extensive tissue damage and even lethality in inflammatory syndromes [20]. On the subcellular level, inflammasomes, such as NOD-like receptor family pyrin domain containing 3 (NLRP3), may be central regulators of early adipose tissue inflammation. Also, inflammasomes are pattern-recognition receptors (PRRs) that assemble into larger structures that control the maturation and secretion of proinflammatory interleukins such as IL-1β. NLRP3 releases the cysteine protease caspase-1, which catalyzes the conversion of procytokines into their mature active forms. The expression of inflammasome NLRP3 components is increased in obesity. Several endogenous stress signals, including glucose, palmitate, cholesterol crystals, islet amyloid peptides and RONS (ROS), can induce inflammasome *in vivo*. Also cellular components of the innate and adaptive immune systems may contribute to adipose tissue dysfunction. The stromal vascular part of adipose tissue contains various types of immune cells, including macrophages. Still, human adipocytes and preadipocytes seem to be able to prime inflammation and attract T-cells independently of macrophages. Moreover, there are more macrophages, but also of proinflammatory T-cells in the subcutaneous adipose tissue of type-2 diabetes patients. T-cells that are infiltrated in adipose tissue may not only attract macrophages, but also skew their differentiation towards the M1 phenotype. In contrast, induction of T-regulatory cells can be beneficial and reduce adipose tissue inflammation and insulin resistance. The anti-inflammatory master switch in adipocyte differentiation, PPAR-γ is helps to control visceral adipose-tissue-resident regulatory T-cells. The gut microbiome may also influence the inflammatory response that is seen in insulin resistance and type-2 diabetes. Gut-derived endotoxins can potentially trigger Toll-like receptors in adipose tissue or on pancreatic β-cells, thus contributing to both insulin resistance and β-cell failure. At the same time, in the liver, portal endotoxemia may contribute to inflammation in hepatic steatosis and be a relevant risk factor for nonalcoholic steatohepatitis (NASH)" [37].

There are two early activators of adipose tissue inflammation in obesity: the complement system and advanced glycation end products (AGEs) [37]. "The complement system is a complex protein network that is best known as being part of the innate immune system. It is produced in the liver, adipose tissue and endothelial cells. Complement activation is under tight control by many circulating and cell-bound inhibitors that prevent unrestrained progression of complement activation once any of three pathways is initiated. They are the classical, the lectin and the alternative pathways. They all converge on complement C3, the central component of the complement system. The alternative pathway is an amplification loop that enhances complement activation once it is initiated by activation by any of the three pathways. All three pathways activate the terminal complement pathway. The complement system is an essential regulator of cell and tissue homeostasis, as well as increasing immunity. Adipose tissue expresses many complement components and regulators. The alternative complement pathway is activated in adipocytes. The expression of proximal components of the classical pathway is altered in the subcutaneous adipose cells of insulin resistant individuals. Much data suggest that complement activation in obese people may help recruit immune cells from the circulation to the adipose tissue. This may enhance the cellular immune response in adipose tissue and thereby contribute to adipose tissue dysfunction. In addition to the proposed effects of complement activation on adipose tissue dysfunction, complement activation may play a role in the progression of β-cell dysfunction in type-2 diabetes. Also, complement system activation leads to endothelial dysfunction. Since the microvasculature comprises approximately 98% of the total vascular surface area, complement activation will inevitably affect microvascular function. With respect to atherosclerosis, complement activation by the classical and perhaps also the lectin pathway may aid in removing apoptotic cells and cellular debris. This will have protective effects in the atherosclerotic plaque, while complement activation beyond C3 may be proatherogenic. Moreover, complement activation may be cause atherothrombosis to develop, since proteases of the coagulation and fibrinolysis systems may activate the complement system, and vice versa" [37].

Dyslipidemia (excess fatty acids and low ratio of LDL to HDL) is another feature of metabolic syndrome. This, together with accumulation of ectopic fat depot, can cause pancreatic lipotoxicity [37]. "Increased concentrations of saturated fatty acids can harm β-cells and produce other proinflammatory cytokines. Moreover, adipose tissue contains mature adipocytes, preadipocytes, endothelial cells, fibroblasts, macrophages and other immune cells. The number of adipose tissue macrophages increases in obesity. Thus, chemoattractants are produced locally and attract monocytes to adipose tissue depots, where they can contribute to inflammation and expansion of adipose tissue. This can lead to adipocyte hypertrophy, local hypoxia and adipocyte apoptosis. This generates signals to recruit macrophages. Hypertrophic adipocytes start to secrete low levels of TNF-α, which stimulate preadipocytes and endothelial cells to produce monocyte chemotactic protein. Inflammatory cytokines (TNF-α and IL-6) can cause insulin resistance" [37].

Remember that obesity is not a disease itself, but is a risk factor for other diseases, such as stroke and cardiovascular diseases [1]. "A stroke is the rapid loss of brain function caused by a lack of proper blood flow, needed to carry oxygen to the brain. Strokes also cause inflammation. There are two main categories of stroke: ischemic and hemorrhagic. About 80% of all strokes are due to ischemia. Strokes can be caused by ischemia-reperfusion, thrombosis, embolism or hemorrhage. Ischemia is a lack of blood flow. It is followed be reperfusion, or renewed flow of the blood to the brain. Ischemia-reperfusion can cause extensive oxidative damage to the brain. The enzyme NADPH oxidase is activated in glial cells and reactive oxygen substances (ROS) are produced. Thrombosis is the formation of a blood clot, which can block the flow of blood at the site in the body in which it is formed. Embolisms, on the other hand, are due to blockage by an object called an embolus (plural emboli) that migrates to another part of the body, where it causes blockage. Hemorrhage is an internal bleeding, or loss of blood from the circulatory system" [1].

High blood pressure is the most important modifiable risk factor that can cause stroke [1]. "Often, blood pressure can be reduced by diet, exercise and medication. All these things can also be used to treat high cholesterol, another risk factor for stroke. Other risk factors, such as old age and diabetes, are not so easily controlled" [1]. Moreover, diabetes is the most common modifiable risk factor for chronic kidney disease, followed by systemic hypertension [49].

Ischemic strokes need to be treated as soon as possible [1]. "It is very important to get a stroke victim to a hospital as soon as possible, so the blockage can be removed and blood can begin to flow properly. Drugs such as aspirin, clopidogrel and dipyridamole can be given to prevent more clots from forming by preventing blood platelets from aggregating. If the patient can get to a hospital soon enough, tPA, or tissue plasminogen activator, can be given. The drug called alteplase is a tPA that is produced by recombinant DNA technology. The complementary DNA (cDNA) from the gene that codes for tPA is inserted into Chinese Hamster Ovary cells (CHO cells), which secrete the tPA into the culture medium. Plasminogen is a serine protease that is made in the liver and released into the circulation. Plasminogen is converted to the enzyme called plasmin in a reaction that is catalyzed by tPA. Plasmin catalyzes the degradation of many proteins in blood plasma, especially those proteins that cause fibrin clots. So, tPA indirectly causes the breakdown of blood clots that can cause stroke. In a recent article, it was reported that the therapeutic window for alteplase (or tPA) can be extended from three hours to 4.5 hours [41]. On the other hand, strokes and heart attacks can be prevented by taking aspirin, which prevents the formation of blood clots. People with heart disease often take aspirin (100 mg), along with a beta-blocker (atenolol, 50 mg), a diuretic (a thiazide, 12.5 mg), an ACE inhibitor (ramipril, 5 mg), and a statin (simvastatin, 20 mg). Thiazides include benzothiadiazine, chlorothiazide and hydrochlorothiazide" [1]. Recently, it was shown that these can be combined in one pill (called a polypill) and it was found to be safe and effective in reducing risk factors for stroke and heart disease [50].

When blood flow to the heart is interrupted, it causes a myocardial infarction, or heart attack [1]. "The most common cause is a blockage in the coronary artery that is usually caused by the rupture of an atherosclerotic plaque. Heart attacks are the leading cause of death in men and women world-wide. Immediate treatment includes oxygen, aspirin and nitroglycerin. Heart attacks can be diagnosed by an electrocardiogram and by the analysis of blood for elevated levels of creatanine kinase-MB and either troponin I or troponin T. Creatinine kinase has two subunits, one from the muscle (M) and one from the brain (B). So, there are three possible isozymes, MM, MB and BB. The heart muscle expresses both MM and MB isozymes. Troponin is a complex of three proteins and it is important in making muscles (including the heart) contract when the intracellular $Ca^{+2}$ concentration increases. Troponin, actin and myosin are the components of muscle fibers. Troponin is the protein to which $Ca^{2+}$ binds. When a person shows symptoms of a heart attack, their blood is analyzed for troponin, which would appear if heart cells have died" [1].

Smoldering inflammation can lead to heart attacks [1]. "It was once thought that the circulatory system was like the plumbing in our homes. If our arteries become clogged with plaques that contain cholesterol, it can cause a heart attack. However, we now know that inflammation comes first. First, adhesion molecules (such as vascular cell adhesion molecule-1, or VCAM-1) are expressed by endothelial cells. This allows leukocytes to attach to the arterial wall. VCAM-1 binds to monocytes and T lymphocytes, which are found in early atherosclerotic plaques. Oxidized lipids, nuclear factor-κB (NF-κB) and TNF-α can also stimulate VCAM-1 expression. These things can be prevented by laminar blood flow, which causes some anti-atherosclerotic mechanisms, such as expression of the natural anti-oxidant, superoxide dismutase and an increase in nitric oxide (NO) synthetase" [1]. There is an increase in the concentration of NO, which causes vasodilation and limits VCAM-1 gene expression [35]. "However, when blood flow is disturbed, lesions can form and monocytes can adhere to the arterial endothelium. Monocytes can then enter the vessel walls, causing a fatty streak to start developing. This is the first stage of atherosclerosis. Monocytes mature into macrophages, which express scavenger receptors and engulf modified lipoproteins. Cholesteryl esters accumulate in the cytoplasm and macrophages become foam cells, which have an increased concentration of lipids. Next, T-lymphocytes of the adaptive immune system enter the arterial cell walls and participate in further inflammation. They are attracted to the arterial cells by chemokines (γ-IP-10, MIG and I-TAC) that are induced by interferon-γ" [35].

The next stage is progression to the formation of plaque, which can begin early in life, when a person is still in their early twenties [35]. "A mixture of cytokines: interleukins (IL-1β, IL-1α, IL-6, IL-18) and tumor necrosis factors (TNF-α, TNF-β) are involved in this process, as are the macrophage colony stimulating factor (MCSF) and the monocyte chemoattractant protein, MCP-1. Next, the plaques can rupture, leading to thrombosis. In healthy arteries, there is a fibrous cap that contains collagen for strength. The strong cap protects blood from contact with the lipid core. When the cap ruptures, a thrombus, or blood clot, can form. Inflammation interferes with the integrity of the collagen matrix by blocking the creation of new collagen fibers and by stimulating the destruction of existing collagen. T-lymphocytes participate in this pro-inflammatory process. They produce CD40, a cell surface protein and IL-1. These biomolecules promote the production of enzymes in macrophages that catalyze the destruction of collagen. The CD40 ligand also stimulates macrophage production of tissue factor VII, which initiates the coagulation cascade. To prevent all these things from happening, one should try to lower the risk factors by not smoking, engaging in regular strenuous cardiovascular exercise and by consuming a healthy diet, with fewer calories, and more fresh fruits and vegetables and other sources of antioxidants" [35].

Also, hyperglycemia and diabetes often lead to heart disease. Vascular homeostasis is disrupted due to endothelial and smooth muscle cell dysfunction [35]. "This causes a further inflammation and a pro-thrombotic state. Elevated levels of RONS (ROS) lead to endothelial dysfunction, superoxide anion ($O_2^-$) production and NO imbalance. Peroxynitrate ($ONOO^-$) easily penetrates and crosses through cell membranes and nitrosylates proteins, which decrease the activity of endothelial NO synthase (eNOS) and antioxidant enzymes. RONS (ROS) also initiate polyol and hexosamine flux, help form AGEs, activate protein kinase C (PKC) and vascular inflammation mediated by NF-κB. Once activated, PKC increases ROS production through NADPH oxidase, which catalyzes the formation of the superoxide anion. Also, the PKC $\beta_2$ isoform catalyzes the phosphorylation of another protein, $p66^{Shc}$, causing it to move to the mitochondria, where it functions as a redox enzyme that generates more ROS and translates oxidative signals into apoptosis. Moreover, PKC leads to higher production of endothelin-1 (ET-1), which favors vasoconstriction and platelet aggregation. In the blood vessel wall, RONS (ROS) initiate vascular inflammation. They up-regulate NF-kB subunit p65 and move it to the cell nucleus, where it activates the transcription of pro-inflammatory genes encoding for monocyte chemoattractant protein-1 (MCP-1), selectins, vascular cell adhesion molecule-1 (VCAM-1), and intracellular cell adhesion molecule-1 (ICAM-1). This helps monocytes stick to the vascular endothelium, leading to the formation of foam cells. Adhesion molecules stay up-regulated by the secretion of IL-1 and TNF-α from active macrophages. This enhances NF-kB signaling in the endothelium and also helps smooth muscle cells grow and proliferate. Also, endothelial dysfunction increases the synthesis of vasoconstrictors and prostenoids. Cyclooxygenase-2 (COX-2) is up-regulated by PKC and this is associated with an increase of thromboxane A2 and a reduction of prostacyclin (PGI2) release. As mentioned previously, RONS (ROS) also increase the level of methylgyoxal and AGEs, which led to protein nitrosylation. This includes OGlcNAcylation at the Akt binding site of the eNOS protein. This leads to reduced eNOS activity and endothelial dysfunction. Also, glycosylation of transcription factors causes up-regulation of inflammatory (TGFα, TGFβ1) and pro-thrombotic genes (plasminogen activator inhibitor-1)" [35].

Insulin resistance, caused by diabetes, develops in not just the heart, but also skeletal muscle, adipose tissue and the liver [35]. "Adipose tissue in particular produces inflammatory mediators and free fatty acids that bind to Toll-like receptors (TLRs) and activate NF-kB. They degrade the inhibitory complex formed between IkBα by IKKβ-kinase. As a result, NF-kB triggers tissue inflammation due to up-regulation of the pro-inflammatory genes IL-6 and TNF-α. Free fatty acid activation of TLRs leads to phosphorylation of the insulin receptor substrate-1 (IRS-1) by c-Jun amino-terminal kinase (JNK) and PKC. This alters its ability to activate the downstream targets PI3-kinase and Akt. This down-regulates the glucose transporter GLUT-4, leading to insulin resistance. Moreover, down-regulation of the PI3-kinase/Akt pathway leads to inhibition of eNOS and decreased NO production. Then, intracellular oxidation of free fatty acids produces ROS. This leads to vascular inflammation, synthesis of AGEs, reduced prostaglandin I2 (PGI2) synthase activity, and PKC activation. As more RONS (ROS) are formed, they can react with NO, reducing its level even more. This activates pro-inflammatory pathways that are activated by increased cytokine production. That is, TNF-α and IL-1 increase NF-kB activity and the expression of adhesion molecules. TNF-α also stimulates the expression of C-reactive protein, which down-regulates eNOS while increasing the production of adhesion molecules and endothelin-1" [35].

However, recent data has indicated that the endothelium is more important than adipocytes in causing obesity-induced insulin resistance [51]. "That is, inflammation and macrophage activation occur primarily in non-adipose tissue in obese people. Also, insulin sensitivity can be imporved in patients with type-2 diabetes by resoring flow-mediated vasodilation. Still, changes in the blood lipid profile (high triglycerides, low HDL cholesterol, increased remnant lipoproteins, elevated apolipoprotein B, ApoB, and elevated LDL) can help cause atherosclerosis. When free fatty acids reach the liver, very low density lipoproteins (VLDLs) are assembled and are solubilized by making more ApoB. VLDLs are processed by cholesteryl ester transfer protein. This allows triglycerides to be transferred to LDLs, which become smaller and denses and, thus, more atherogenic. As a result, atherogenic dyslipidemia is a reliable predictor of cardiovascular risk and its pharmacological modulation reduces vascular events in subjects with type-2 diabetes and metabolic syndrome" [54].

Harmful coronary events in patients with insulin resistance are triggered by being in a prothrombotic state [51]. "Under healthy conditions, insulin inhibits platelet aggregation and thrombosis by inhibiting tissue factor (TF) and enhancing fibrinolytic action by changing the concentration of plasminogen activator inhibitor-1 (PAI-1). In contrast, insulin resistance facilitates atherothrombosis by increasing the biosyntheses of PAI-1 and fibrinogen while reducing the production of tissue plasminogen activator. In platelets, a lack of insulin leads to a down-regulation of the IRS-1/Akt pathway. This causes $Ca^{2+}$ to accumulate. Moreover, platelet reactivity and excretion of tromboxane metabolites are increased in insulin resistant obese patients and this is reversed by losing weight or by a 3-week treatment with pioglitazone" [51].

More recently, researchers have found that microRNAs (miRNAs) are involved in the pathogenesis of vascular damage caused by hyperglycemia [51]. "That is, miRs involved in angiogenesis, vascular repair, and endothelial homeostasis are dysregulated in diabetic patients. For example, in endothelial cells exposed to excess glucose, miR-320 is highly expressed. It then targets several angiogenic factors and their receptors, including vascular endothelial growth factor and insulin like growth factor-1 (IGF-1). This is associated with decreased cell proliferation and migration. Hyperglycemia also increases the expression of miR-221, a regulator of angiogenesis that targets the c-kit receptor, which is responsible for migration and homing of endothelial progenitor cells (EPCs). Also, miR-221 and 222 were mediate AGE-induced vascular damage. Another study showed that miR-503 is critically involved in hyperglycemia-induced endothelial dysfunction in diabetic mice and is up-regulated in ischemic limb muscles of diabetic subjects. Also, profiling of plasma miRs showed that miR-126 was down-regulated in a cohort of diabetic patients. This may be partially responsible for impaired vascular repair capacities in diabetes. Finally, miR-126 expression was reduced in endothelial progenitor cells that in diabetics" [51].

## CANCER

Cancer is the rapid, unregulated and pathological growth (proliferation) of abnormal cells [1]. "Even when cancers occur in the same part of the body, they can be very different diseases. When tumors become malignant, their threat depends on their ability to modify surrounding cells to form new blood vessels (angiogenesis) and other supporting cells" [1].

Tyrosine kinases regulate many cellular processes which can contribute to cancer development and progression [1]. "Oncogenes code for oncoproteins, which are upregulated in cancer. KRAS is the oncoprotein that is most commonly activated in human cancer. *RAS* is one of the most commonly mutated genes in human cancers. Another important oncogene is *PI3K*, which codes for the enzyme PI3K (phosphoinositide 3-kinase). The enzyme PTEN (phosphatase and tensin homolog) catalyzes the opposite reaction, so it is a tumor suppressor and the gene coding for it is downregulated in cancer. Human epidermal growth factor (hEGF, or HER), vascular endothelial growth factor, or VEGF, and the PI3K/Akt/mTOR (mammalian target of rapamycin) survival pathway are all important therapeutic target in many cancers" [1].

The hallmarks of cancer are inflammation, chromosomal defects, genetic instability, escape from immunosurveillance, limitless proliferative potential, self-sufficiency in growth signals, insensitivity to antigrowth signals, evasion of apoptosis, sustained angiogenesis, tissue invasion and metastasis [1]. "Smoldering inflammation can lead to cancer and other diseases. Reactive oxygen substances, such as the hydroxyl radical, produced by poorly liganded iron, can cause oxidative damage to DNA, proteins and lipids, affecting the genome, epigenome, proteome and lipidome. Inflammatory cytokines can be produced. Moreover, mutations accumulate as we age, eventually leading to new cell surface antigens being made. These things can change a tumor's interaction with the immune system, enabling it to escape immunosurveillance. Then, it develops limitless proliferative potential as it becomes self sufficient in producing the growth signals it needs and becomes insensitive to antigrowth signals. Finally, a tumor becomes deadly when it builds new blood vessels and metastasizes into the lymph nodes and other organs" [1].

Most cells in our body are broken down and re-made [1, 52]. "When a cell is damaged or just reaches the end of its normal life, it usually undergoes programmed cell death, also known as apoptosis. There is an intrinsic and an extrinsic pathway to apoptosis. The extrinsic pathway begins outside the cell, while the intrinsic pathway begins inside the cell. In both pathways, a class of cysteine proteaseses called caspases catalyzes reactions that dismantle dead and dying cells. When this apoptosis is disrupted, cancer can begin. Abnormal cells should die, but they don't. The mitochondria, which regulate apoptosis in healthy cells, become defective. Not only do they not initiate apoptosis, but their metabolism changes. Despite the ready availability of $O_2$, glycolysis occurs as if it was in an anaerobic environment and the end product is lactate, not pyruvate. This is called the Warburg effect, after the scientist who discovered it. At first, this was thought to be a cause of cancer, but we now know that it is an effect, or a symptom" [1]. Recent data has shown that the loss of the tumor suppressor SIRT6 may be the cause of the Warburg effect [53].

On the other hand, when a cell suffers oxidative damage in healthy tissues it can induce apoptosis" [1]. "In the extrinsic pathway, apoptosis is initiated by the binding of death ligands to death receptors on the cell membrane. Death receptors belong to the tumor necrosis factor receptor (TNFR) gene superfamily that includes TNFR-1, Fas/CD95, and the TRAIL (tumor necrosis factor-related apoptosis inducing ligand) receptors DR-4 and DR-5. They cause apoptosis mediated by caspase 8, a cysteine-aspartate protease. On the other hand, the intrinsic pathway begins when there is damage within the cell. It can be caused by oncogenes, direct DNA damage, oxidative stress and starvation. The cytoplasmic proteins, BAX and BID, bind to the outer mitochondrial membrane upon receiving appropriate stress signals. BAX and BID interact with a protein called BAK that is in the mitochondria. This causes the release of cytochrome c and other mitochondrial proapoptotic factors into the cytosol. Once in the cytosol, cytochrome c binds to apoptotic protease activating factor-1 (Apaf-1) which then forms an apoptosome that triggers the activation of the initiator procaspase-9. That is, caspases are regulated at a post-transcriptional level, so they can be rapidly activated when needed. They are first biosynthesized as inactive pro-caspases. They are only activated when needed. Activated caspase-9 initiates a caspase cascade, ultimately resulting in cell death. Tumor suppressor p53 protein is a sensor of cellular stress and is a critical activator of the intrinsic pathway. It initiates apoptosis by activating the transcription of proapoptotic BAX family members and repressing antiapoptotic Bcl2 proteins" [1].

On the other hand, p53 is not involved in the induction of apoptosis in the extrinsic pathway [1]. "Moreover, the orderly progression of the cell cycle from one stage to the next is controlled by checkpoints that prevent unwanted mitosis in response to DNA damage. Following DNA damage, the cell cycle can be arrested to enable DNA to be repaired. If the damage can't be repaired, apoptotic pathways are activated. So, there are genome stability genes that code for proteins that regulate the cell cycle or DNA repair mechanisms" [1].

The key regulators of cell cycle progression are the cyclin-dependent kinases (CDKs), cyclins, and CDK inhibitors [1]. "The cyclin/CDK complexes promote cell cycle progression. On the other hand, inhibitors of the CDK protein can stop or arrest the cell cycle. Moreover, stem cells can play an important role in cancer" [1].

Tissue-specific stem cells can be transformed into cancer stem cells, which can differentiate to form any of the many types of cancer cells that can exist in a tumor [1]. "So, cancer is the rapid, unregulated and pathological growth (proliferation) of abnormal cells" [16]. That is, some cancers are organized into a hierarchy of subpopulations of tumorigenic cancer stem cells and their non-tumorigenic progeny. In these cases, cancer stem cells are thought to drive tumor growth and disease progression, perhaps through resistance to therapy and metastasis [54]. "However, even in cancers that clearly have a hierarchy of tumorigenic and non-tumorigenic cells, other factors are important. Other sources of heterogeneity include clonal evolution, heterogeneity in the micro-environment and reversible changes in properties of cancer cells that can occur independently of hierarchical organization" [54]. Nowell's theory of clonal evolution proposes that cancers arise from a single cell of origin, develop genomic instability during replication and then the most aggressive clones are enriched by metastasis and the eradication of competing clones that are killed by cancer treatment [55].

Often, cancer is thought to be a process of asexual evolution driven by genomic and genetic instability [56, 57]. "There is much variation in the genotypes and phenotypes between tumors and within individual tumors, caused by genetic instability. Genetically distinct subclonal populations of cancer cells can emerge through intercellular genetic variation. This can cause selective outgrowth of clones of cells that have a phenotypic advantage within the micro-environment of the tumor. The new micro-environment at metastatic sites can help explain genetic differences between metastatic and primary sites. Moreover, there is much heterogeneity in both the patterns and extent of genomic instability in different types of tumors" [56].

Mutation, selection and adaptation were once thought to occur primarily within, and to a lesser extent, outside the primary tumor [1]. "However, cancer cells that remain after surgery can have extreme genomic heterogeneity before the metastasis begins. Later, this heterogeneity decreases due to selective clonal expansion, suggesting that the remaining cells had yet to acquire many of the important traits that fully malignant cells have. Trying to stop the cells' progression outside the primary tumor offers new challenges and opportunities for diagnosis and adjuvant therapies" [1].

This is important because cancer is the second leading cause of death in the USA, after cardiovascular disease [1]. "There are over 100 types of clinically distinct cancers and many subsets of each type. So, cancer is actually many different diseases that occur in various parts of the body, such as the skin, stomach, mouth, throat, lungs, intestines, breasts, testicles, kidneys, prostate, pancreas, liver, ovary, bladder, blood cells and bone. Even when cancers occur in the same part of the body, they can be very different diseases. For example, the most common form of skin cancer is not very dangerous, but the rarer form, melanoma, is often fatal. Cancer in other parts of the body can have a wide range of prognoses. Cancers can be classified according to their state of advancement, rate of growth and lethality. In the beginning, abnormalities can be benign. Malignant tumors are able to form vascularized colonies, called metastases, in different parts of the body. Benign growths can't do this. They are contained and are relatively harmless, but still threatening, since the abnormal cells do have the potential to become malignant. The difference between benign and malignant lesions is associated with the emergence of genome instability" [1]. One main feature or hallmark of cancer is an ongoing production of genomic diversity, which enables gene networks to self-organize and individual cells to progress towards metastasis in a process of clonal evolution [57]. The process of self-organization can be described in network theory as the formation of attractor states [1].

Once tumors become malignant, their threat depends on their ability to modify surrounding cells to form new blood vessels (angiogenesis) and other supporting cells [58]. "There can be cancer cells that circulate in the blood (circulating tumor cells, or CTCs) and cancer cells that disseminate into other organs (disseminated cancer cells, DCCs). During successful dissemination, tumor cells invade the surrounding tissue of the primary tumor, intravasate into blood and lymphatic vessels, translocate to distant tissues, extravasate, adapt to the new microenvironment, and eventually seed, proliferate, and colonize to form metastases. Because dissemination occurs mostly through the blood, CTCs that have been shed into the vasculature and may be on their way to potential metastatic sites are very important. Because blood collection is simple and minimally invasive, CTCs could be used as excellent markers for predicting disease progression and survival. Identification of CTCs may also be able to guide therapeutic management and effectiveness, even in the absence of detectable metastases, and offer insights into mechanisms of drug resistance. However, CTCs have a diameter that is three to four times as large as the bores of capillaries in organs, so they can become trapped there. CTCs can be detected in the blood, but only extremely small and/or plastic CTCs can keep circulating. Also, some CTCs circulate as microclusters that may become stuck in capillaries. So, it is unclear whether actual CTCs are truly the source of metastases. Moreover, tumors can spread through the lymphatic system. So, single cell assays can be especially informative. For example, CTCs that have the human epidermal growth factor receptor 2 (HER2) on them have been found in HER2-negative breast cancers" [58].

So, cancer cells move to other parts of the body in a process called metastasis and become disseminated metastatic tumor cells (DTCs). Once DCCs become metastatic, their genomes become quite similar, indicating that an aggressive clone has expanded. So, most forms of breast cancer can almost always be cured if they are caught before they move (metastasize) into other tissues [10]. Moreover, metastasis accounts for the majority of cancer patients' deaths. So, drugs that target angiogenesis continue to be developed and some are already approved [59]. These drugs work by inhibiting angiogenesis: Bevacizumab (Avastin®), cetuximab (Erbitux ®), panitumumab (Vectibix ®), and trastuzumab (Herceptin ®) [59].

So, a tumor is a collection of diverse cells, not just cancer cells that have carcinogenic mutations [1]. "Within tissues, there are epithelial cells that are surrounded by a connective framework, called stroma. The stroma consists of fibroblasts, endothelial cells, immune cells and an extracellular matrix (ECM). The ECM provides cells in tissues with information that tells them to survive. In healthy cells, the stroma is the main barrier against tumorigenesis. However, in the presence of transformed tumor cells, changes occur that support cancer progression. Fibroblasts are recruited, immune cells migrate and the ECM is remodeled and eventually vascular networks develop. Healthy fibroblasts maintain the structural framework in tissues. Quiescent fibroblasts respond differentially to damage and are activated to support repair. Although normal fibroblasts typically suppress tumor formation, cancer-associated fibroblasts (CAFs) can promote tumorigenesis. Compared with normal tissue fibroblasts, CAFs have increased proliferation, enhanced extracellular matrix production and secrete unique cytokines such as stromal cell-derived factor 1 (SDF1), vascular endothelial growth factor (VEGF), platelet-derived growth factor (PDGF) and hepatocyte growth factor (HGF)" [1]. Other mesenchyme-derived cell types, such as adipocytes, can also contribute to the growth and progression of tumors [61]. Differences in fibroblast behavior and response can cause extensive tissue remodeling that is mediated by increased expression of proteolytic enzymes such as matrix metalloproteinases. Also, pathogenic angiogenesis can occur by liberating pro-angiogenic factors within the ECM.

The tumor vasculature network is dynamic and can limit the growth of tumors [61]. "They are made by the formation of new vessels (angiogenesis). So, existing vessels within tissues, or the recruitment and differentiation of endothelial precursors from bone marrow (vasculogenesis) can contribute to vascular heterogeneity in and among tumors. Uneven vascularization and differences in vascular maturity, when combined with a lack of drainage due to poor lymphatic vessel coverage will cause complex topography and variable interstitial pressure within tumors. When tumor vasculatures are poorly organized and don't function properly, it produces areas of hypoxia and a limited supply of nutrients. A gradient develops that is based on different distances from the vascular beds. It is crucial that drugs are distributed to all cells in the tumor. These vascular networks can generate distinct micro-environments within the tumor and contribute to inter- and intratumor heterogeneity, and ultimately influence the clinical outcome. Microvessel density may also be a significant prognostic factor for indicating a poor outcome in non-small-cell lung cancer (NSCLC) as well as colorectal and breast cancers. Moreover, elevated expression of the predominant pro-angiogenic ligand VEGFA is associated with a worse prognosis than those that have a relatively low expression of VEGFA in metastatic colorectal, lung and renal cell cancers" [61].

As signals come from the ECM through cellular adhesion proteins, biochemical pathways can be activated that ensure the survival of cell [1]. "When such cells lose contact with the ECM, different pathways are activated and the cells undergo a type of programmed cell death called anoikis. However, in advanced stages of cancer, this mechanism fails. Cancer cells that break free from the ECM don't die. They migrate (metastasize) into other parts of the body, where they can be especially deadly" [1].

Also, a healthy immune system can recognize and protect tissues from infections and damage [1]. "However, both the innate and adaptive immune systems can promote and prevent tumor growth. Although the immune system has the ability to mount anti-tumor responses, immune suppression mechanisms can prevent this process. That is, many types of cancer cells can suppress the immune system. Cancer cells can escape detection by decreasing the expression of important antigenic proteins, making them invisible to cytotoxic T cells. They can also secrete proteins that inhibit effector T cells and enhance the production of regulatory T cells that suppress immune responses" [1]. Some melanomas can reorganize their surrounding cells (stroma) to form structures that are similar to those of lymphoid tissue in the immune system. They can recruit and maintain immune cells that promote tolerance and tumor progression [62].

Besides, the activation of T-cells involves both stimulatory and inhibitory checkpoint signals that finely tune responses in order to prevent excessive damage and autoimmunity [61]. "One way to usurp the activation of cytotoxic T-cells in tumors is by continuously engaging inhibitory receptors on T-cells, such as cytotoxic T-lymphocyte-associated antigen-4 (CTLA-4) and programmed death 1 (PD1) by upregulating their ligands. An indirect way of preventing antitumor T-cell responses is to generate an immunosuppressive environment. Myeloid-derived suppressor cells (MDSCs, which includes neutrophils, immature dendritic cells, monocytes and early myeloid progenitors) were expanded upon upon tumor implantation. This implies that there is tumor-initiated endocrine communication with the immune system and secretion of chemokines by tumor and cancer associated fibroblasts. This includes granulocyte–macrophage colony stimulating factor, GM-CSF; or granulocyte colony stimulating factor, G-CSF. The recruitment of immunosuppressive myeloid lineages to the tumor not only suppresses adaptive immunity, but also fosters angiogenesis through the secretion of VEGFA, basic fibroblast growth factor (bFGF) and transforming growth factor β (TGF-β). MDSCs also inhibit natural killer cell function and expand the immunosuppressive regulatory T-cell population. Also, MDSCs can directly inhibit the expansion activation and migration of effector T-cells by changing the environment. B cells can suppress and support T-cell function, resulting in differential effects on tumorigenesis. However, B cells can promote tumor progression by increasing inflammation. The recruitment of mast cell has been implicated in tumorigenesis and angiogenesis. Tumor-associated macrophages (TAMs) can also drastically affect tumor progression, depending on their polarization" [61].

Moreover, different subtypes of immune cells have differential prognostic value [61]. "For example, in melanoma, colorectal and breast cancer, there is a strong correlation between high T-cell occupancy and a good clinical outcome. However, other types of T-cells, such as regulatory T cells, and helper T cells ($T_H1$ and $T_H17$) are associated with poor or good prognosis depending on the type of cancer. In breast cancer, assessment of pro- and antitumor immunity showed that a high ratio of macrophages to cytotoxic T-cells was correlated with a worse outcome than those with a low ratio. That is, a stromal signature derived from microdissected breast tumor biopsies can help predicted the prognosis, regardless of the tumor subtype. Angiogenesis, hypoxia and macrophage-mediated immune suppression signatures have been associated with poor prognosis. Also, B-cell content has been associated with a good prognosis in ovarian and breast cancers. This provides a strong argument for considering the content of the immune network to be a major contributor to tumor development [61].

The stroma influences not only cancer cells, but also cells that are present in a wound. The stroma can support the growth and spread of cancer, and it can help wounds heal. Scientists have examined the DNA from stroma cells surrounding tumors and found 26 genes that are called the stroma-derived prognostic predictor, SDPP [63]. Analysis of these genes can be used to help establish the proper therapy for a patient. The sequence of bases in these genes varies with the prognosis of breast cancer, because once a cancer has metastasized into other parts of the body, it is often fatal. Thus, it is very important to diagnose cancer as early as possible. The interactions between tumors, stroma and the microenvironment are becoming key considerations when developing new anticancer drugs [64].

Recent data has helped show how obesity can facilitate cancer [65]. That is, stromal progenitor cells from endogenous adipose tissue contribute to pericytes and adipocytes that populate the tumor microenvironment. So, obesity facilitates tumor growth in mice irrespective of their diet, suggesting a direct effect of excess white adipose tissue on cancer growth [65].

Still, the first step in starting cancer is usually one or more mutations [1]. "For example, carcinogens in tobacco products can bind to DNA and react with it, producing mutations. DNA can also be damaged by UV radiation and by reactive oxygen substances, or ROS. Thus, many people feel that cancer can be prevented by consuming antioxidants, which destroy ROS and bind to $Fe^{2+}$, keeping it from forming the hydroxyl radical – an especially damaging ROS. Sometimes, mutations are caused by errors in copying DNA, prior to cell division. However, cells have enzymes that can proofread the copied DNA and catalyze the correction of mistakes or bases that are mis-matched. Such enzymes are often important in regulating cell growth when the cell is healthy. They control cell cycle check points and help determine whether a cell should live or die. When these DNA repair enzymes are mutated, the cells can become immortalized cancer cells. As a person gets older, the number of uncorrected mutations can accumulate, so age is an important factor in cancer" [1].

Another factor is that the immune system becomes weaker in older people. Our immune systems produce cells that can kill some cancer cells before they develop into advanced stages. However, many cancer cells avoid attack by the immune system because they carry self-antigens on their outer membrane surface [16]. "Another age-related factor is that normal human cells can only copy themselves about 60 times. This is partly due to telomeres, or tandem repeats of the following DNA base sequence: TTAGGG, located at the ends of chromosomes. These telomeres become shorter after each cell division, until they are so short that the cell can't divide any more. Cancer cells usually acquire the ability to regenerate telomeres, enabling them to copy themselves many times" [16].

When a healthy cell copies itself in mitosis, it goes through a cell cycle [16]. The first stage of the cycle is the resting state, called $G_0$. If there are sufficient nutrients, a cell can enter a growth stage, or $G_1$. There is a restriction point (called R), at which the cell will decide whether to remain in $G_1$ or proceed to the stage of DNA synthesis (S). There are proteins called R-point gatekeepers, which allow passage into the next stage only if the proteins are phosphorylated. The level of phosphorylation is controlled by signals arriving from outside the cell. These signals activate a tyrosine kinase, which catalyzes the phosphorylation of tyrosine residues. The human genome codes for 90 proteins with tyrosine kinase domains [66] and for 518 different kinases [16]. Much research is being done to understand them better [16].

Tyrosine kinase-linked receptors (also called receptor tyrosine kinases, or RTKs) are receptors that are linked to tyrosine kinases [1]. "These receptors are also enzymes that catalyze the addition of a phosphate to tyrosine residues in proteins. RTKs are receptors for many polypeptide growth factors, cytokoines and hormones. Of the 90 unique tyrosine kinase genes identified in the human genome, 58 encode receptor tyrosine kinase proteins. RTKs have been shown to be not only key regulators of normal cellular processes but also to have a critical role in the development and progression of many types of cancer. Most RTKs have only one subunit, but some, such as the insulin receptor, have more than one. Each monomer has a single domain that spans the membrane, an extracellular N-terminal region and an intracellular C-terminal region. The extracellular N-terminal region is composed of a very large domain which binds to extracellular ligands. The intracellular C-terminal region comprises domains responsible for the kinase activity of these receptors. When a growth factor binds to the extracellular domain of an RTK, its dimerization is triggered along with other adjacent RTKs. This leads to a rapid activation of the protein's cytoplasmic kinase domains, the first substrate for these domains being the receptor itself. The activated receptor then becomes autophosphorylated on multiple specific intracellular tyrosines. The phosphorylation of specific tyrosines within the activated receptor creates binding sites for proteins that contain Src homology-2 (SH2) and phosphotyrosine binding (PTB) domains. Specific proteins containing these domains include Src and phospholipase Cγ. The phosphorylation and activation of these two proteins leads to the initiation of signal transduction pathways. There are other proteins that interact with the activated receptor act as adaptor proteins and have no intrinsic enzymatic activity of their own. These adaptor proteins link RTK activation to downstream pathways, such as the MAP kinase cascade" [1].

Tyrosine kinases regulate many cellular processes which can contribute to cancer development and progression [66]. "These cellular processes include cell growth, differentiation, migration and apoptosis. There are several anti-cancer drugs that inhibit tyrosine kinases in some patients. These include imatinib (for chronic myelogenous leukemia), tratuzumab (for breast cancer) and three drugs (cetuximab, erlotinib and gefitinib) that inhibit the tyrosine kinase enzyme called the epidermal growth factor receptor (EGFR) and are used to treat cancers of epithelial cells, such as colorectal, head and neck cancers" [66]. Cetuximab is a chimeric mAb, while erlotinib is a reversible inhibitor of EGFR, which catalyzes the phosphorylation of some of its own tyrosine residues, which activates downstream signaling cascades, including MAPK.

However, cancer cells often mutate when treated with tyrosine kinase inhibitors and survive [1]. "They become dependent on one or just a few oncogenes to maintain the malignant phenotype. Two of them are PUMA and BIM. PUMA is a member of the BCL2 family of proteins. It is activated upon inhibition of the PI3K-Akt pathway in both HER2-amplified breast cancers and EGFR mutant lung cancers. BIM is also a BCL2 family member that plays a central role in gefitinib and erlotinib-induced apoptosis in EGFR mutant lung cancers and in lapatinib-induced apoptosis in HER2-amplified breast cancers" [1]. There are distinct subsets of cancers that depend on specific driver mutations for survival, a phenomenon called oncogene addiction [67].

In addition, some viruses produce cancer by inserting (transfecting) their genes into those of the host [16]. The cancer-causing oncogenes can code for proteins that propagate signaling cascades which are not properly regulated. For example, there are viruses that cause sarcoma, or cancer of connective tissues, such as bone or muscle. One of these is called the Harvey virus and it converts the normal, healthy *RAS* gene into an oncogene (*HRAS*) and another, called the Kirsten virus converts *RAS* into *KRAS*. KRAS is the oncoprotein that is most commonly activated in human cancer [67]. "However, oncogenic KRAS and its known downstream effectors have thus far presented intractable targets for antineoplastic drugs" [67]. Still, proteins and pathways associated with oncogenic KRAS have been identified, so they provide new therapeutic targets [67]. Each virus was named after the person who discovered it. *RAS* is one of the most commonly mutated genes in human cancers. It is found in about 20% of them, but it has proved impossible to effectively target with small molecule inhibitors. The *HRAS* and *KRAS* oncogenes code for a Ras oncoprotein, which is a mutant tyrosine kinase that is always switched on. The Ras oncoprotein affects the transcription of many genes. Many R point proteins are hyperphosphorylated. Several proteins are synthesized and apoptosis is inhibited [16]. *HRAS* and *KRAS* oncogenes don't do this directly, but instead act through further (downstream) pathways, including one that involves NF-κB. Hopefully, the deadly effects of *HRAS* and *KRAS* oncogenes can be stopped by inhibiting NF-κB.

The oncogene *TP53* codes for the oncoprotein p53 that is also a target for breast cancer therapy [1]. "It is a hub in the scale-free cellular network. A scale-free network is dominated by a few well-connected nodes, called hubs. Most nodes in the network have a few connections, but a small number of nodes have a seemingly unlimited number of connections. It regulates the cell cycle and helps prevent cancer. Under normal conditions, there is not much p53 in the cell, but when cells are stressed, p53 becomes activated and stabilized. It can be activated by post-transcriptional modifications. When it is mutated, p53 becomes an oncoprotein and can help cause cancer. It is the most commonly found oncoprotein in human cancers. It can activate DNA repair, stop the growth of cells, and can induce apoptosis. So, when a cell is exposed to mutagens the DNA can be damaged. The normal p53 protein can either cause the DNA to be repaired, or it can cause the defective cell to die (apoptosis), preventing its growth and differentiation into cancer cells" [1]. However, when mutated, p53 can cause cancer. So, there is a cancer vaccine undergoing clinical trials, and its target is p53 [68].

Since the p53 protein is a hub in the network of cellular metabolism, it interacts with many other proteins [1]. "It regulates the expression of a wide variety of genes in response to genotoxic or cellular stress. The genes are involved in apoptosis, growth arrest, inhibition of cell cycle progression, differentiation and accelerated DNA repair or senescence. Activated p53 can initiate apoptosis and stop cell proliferation. When p53 is mutated and no longer active, uncontrolled cell proliferation, or cancer, can occur. Progression of a healthy cell through the cell cycle is regulated by many different proteins at several check points.

Information from other proteins will affect the check point proteins. The p53 protein is tetrameric. It interacts with many other biopolymers in cells. Parts of it have flexible structures, enabling it to bind to different proteins and different regions of DNA" [1]. It is a transcription factor that can activate genes to produce proteins which either halt cell cycle progression or induce apoptosis [1, 16]. Thus, normal (or wild-type) p53 is able to sense abnormal cell function, making it an effective tumor suppressor [1]. "It can induce or suppress the expression of several genes that code for proteins involved in apoptosis, cell cycle control and senescence. It acts as a guardian of the genome when it is acting properly. However, a mutated form of p53 is the most frequently mutated oncoprotein found in human cancers. Transcription of the *TP53* gene is controlled by negative regulators called *MDM2* and *MDM4*. Even in cancer cells that have the normal, wild-type p53 gene, the activity of the p53 protein is often inhibited by MDM2, another oncoprotein. Deregultaion of the balance (homeostasis) of the ratio of p53/MDM2 can lead to the malignant transformation of cells. Small molecules or potential drugs are being tested for their ability to prevent MDM2-p53 interaction" [1].

The activity of the p53 protein can also be modified by phosphorylation, methylation, acetylation, glycosylation, ribosylation, summoylation, ubiquination and by interaction with other proteins, such as Mdm2 and MdmX, which are ubiquitin ligases that can also bind to the trans-activation domain of the tetrameric p53 protein [69]. "There are other proteins that can interact with Mdm2 and affect the stability and activity of p53. For many years, phosphorylation was thought to be the first step in stabilizing p53 when a cell is under stress. Phosphorylations at Ser395 and Tyr394 do inhibit Mdm2, while phosphorylations at Ser166 and Ser186 activate it. Moreover, cycles of phosphorylation-dephosphorylation and acetylation-deacetylation of Mdm2 affect its ability to inhibit p53. So, drugs that target the interactions between Mdm2 and p53 have been developed and have started clinical testing" [69].

In the second step, modified p53 binds to target genes through the conserved central, core domain, while the C-terminal region was thought to be a negative modulator that had to be modified to allow sequence-specific DNA binding [69]. "The modifications target and modify the C-terminal region of p53, thus affecting the ability of p53 to bind DNA. However, recent research indicated that both the core DNA-binding domain and the C-terminal domain of p53 possess DNA-binding activities. The core DNA-binding domain provides sequence specificity while the C-terminal domain recognizes structural features of target DNA" [69].

Acetylation is an important link between p53 and histones that regulate transcription [69]. "The same acetyl transferases, CBP/p300, Tip60 and Mof catalyze the acetylation of p53 and histones. Mof is also a key regulator of the embryonic stem cell transcriptional network, as it primes genes for developmental programs. However, ubiquitination and acetylation are mutually exclusive. Competition between these modifications may affect the stability of p53. That is, p53 acetylation levels are markedly enhanced in response to stress, promoting p53 stabilization and activation, while ubiquination targets p53 for proteasomal degradation" [69].

So, oncogenes code for oncoproteins, which are upregulated in cancer. Before the protein is mutated, it is called a proto-oncoprotein. An important oncoprotein frequently found in some human cancers in c-Myc. As a proto-oncogene, c-Myc can regulate the expression of 15% of all genes, including genes involved in cell division, cell growth, and apoptosis [70]. "It exerts its effects on transcriptional targets through various mechanisms - there are positive effects from recruitment of histone-modifying enzymes, general transcriptional machinery, and chromatin-remodeling complexes and negative effects from recruitment of DNA methyltransferases. Often c-Myc drives cell proliferation and inhibits differentiation, as in mouse embryonic stem cells. Surprisingly, c-Myc is different in human embryonic stem cells where it induces apoptosis and differentiation [70].

Since it has such an important role and controls so many genes, the presence of the Myc protein in cells must be carefully controlled [1]. "Thus, the transcription of the gene coding for c-Myc is tightly regulated in healthy cells. It is expressed only when cells are dividing. On the other hand, once mutated, the c-Myc gene is over-expressed, causing too much c-Myc protein to be made" [1]. The c-Myc protein or the c-myc gene is overexpressed in a wide variety of human cancers with 80% of breast cancers, 70% of colon cancer, 90% of gynecological cancers, 50% of hepatocellular carcinomas and a variety of hematological tumors possessing abnormal myc expression [71]. On the basis of these frequencies, it is estimated that approximately 100,000 US cancer deaths per year are associated with changes in the c-myc gene or its expression" [71].

The c-Myc protein is a transcription factor that can control the transcription of other genes. Myc forms a complex with another protein, called Max [71]. "When the Myc-Max complex binds to DNA, it can activate transcription. In cancer, though, the c-Myc gene is amplified, so there are many copies of it. Over 200 copies per cell of N-myc can be found in neuroblastomas, and over 50 copies per cell of c-myc, N-myc, or L-myc may be found in small cell lung cancers" [71]. "A highly regulated cell cycle permits cells to repair DNA damage before replicating, thus promoting genomic fidelity. Inappropriate cell cycle proliferation can lead to genomic instability, resulting in new mutations and abnormal chromosome number and structure. So, c-Myc overexpression, even transiently, can induce genomic instability that is characterized by gene amplification, aneuploidy and polyploidy. Other studies suggest that c-Myc induces the production of reactive oxygen species (ROS) by mitochondria, leading to DNA damage and genomic instability" [71].

The Myc-Max basic-helix-loop-helix leucine zipper heterodimer binds a target DNA site, called the E-box [71]. "c-Myc regulates downstream target genes activating cell cycle regulation, apoptosis, or inhibition of cell adhesion. Examples of c-Myc target genes associated with different cellular functions include cell cycle regulation (*p19 Arf, cyclin D, cyclin E, CDK4* and *CUL1* genes), apoptosis (*ODC, LDHA, Bax, FasL* and *PRXIII* genes), metabolism (genes coding for enolase, LDHA, PRXIII, ODC, CAD, NPM, Nucleolin, and JTV-1), cell adhesion (genes coding for collagen, fibronectin and integrin) and cellular differentiation" [71].

The *c-Myc, MYCN* and *L-Myc* genes encode transcription factors that play essential roles in cell proliferation, cell growth, differentiation, and apoptosis [72]. "Myc is a key regulator of many biological activities including cell growth and division (regulation of chromatin modification and components of the biosynthetic network); cell-cycle progression (modulation of cyclins, cyclin dependent kinases, cyclin-dependent kinase inhibitors and phosphatases); apoptosis (p53 dependent or independent mechanisms); cell differentiation (down regulation of growth arrest genes); cell metabolism (glycolysis, amino acid biosynthesis and transport, synthesis of macromolecules and DNA metabolism); angiogenesis (upregulation of VEGF); cell adhesion and motility (control of expression of integrins). Deregulation of Myc may result in apoptosis, genomic instability, uncontrolled cell proliferation, escape from immune surveillance, growth factor independence, and immortalization" [72].

So, amplification of the c-Myc gene can cause cancer chromosomal instability (CIN). This causes an increased number of chromosomes and intratumor heterogeneity. CIN is observed in most solid tumors and is can result in poor prognosis and drug resistance [73]. It causes an increased rate of change of chromosome number and structure and generates intratumor heterogeneity. CIN is observed in most solid tumors and is associated with both poor prognosis and drug resistance [73]. Also, stress can mix chromosomes and cause segregation errors during cell division [74]. "Even though cancer has long been thought to be a disease that cancer occurs after a series of mutations that eventually produce a cell that is well equipped for unrestricted proliferation in the proper environment. Proliferation of this cell would be expected to produce tumors containing cells with this same genome. But in reality, the cells in solid tumors have tremendous genetic variation, or intratumor heterogeneity. Understanding the molecular basis of this heterogeneity may help to understand the cellular pathways that promote tumor formation and progression, and help develop specific anticancer therapies. The genetic variation can be caused by minor DNA mutations and CIN. Structural CIN produces large chromosomal rearrangements, which may be caused by improper repair of damaged DNA. Numerical CIN produces abnormal numbers of chromosomes (called aneuploidy) and has been attributed to various defects that happen during cell division. Replication stress could promote a rapid genetic drift within a tumor, through connected rounds of changes in the structure and the number of chromosomes. Suppressing replicative stress in tumors could inhibit this genetic drift and might prevent the acquisition of resistance to therapies. Such resistance can arise when some cells of a tumor acquire mutations that confer resistance and give them a proliferative advantage. Combining therapies that target replication stress with conventional therapies could help to treat CIN+ tumors" [74].

Another important oncogene is *PI3K*, which codes for the enzyme PI3K (phosphoinositide 3-kinase) [1]. "It catalyzes the phosphorylation of phosphatidyl inositol. Like other oncogenes, *PI3K* is upregulated in cancers. On the other hand, the enzyme PTEN (phosphatase and tensin homolog) catalyzes the opposite reaction, so it is a tumor suppressor and the gene coding for it is downregulated in cancer. PTEN normally blocks a PI3K (phosphoinositide 3-kinase) survival pathway, which allows cells to ignore signals that tell them to commit suicide (apoptosis). One form or isozyme of PI3K catalyzes the addition of a phosphate to phosphatidyl inositol 4,5-bisphosphate ($PIP_2$), to form phosphatidyl inositol 3,4,5-trisphosphate ($PIP_3$). The PTEN protein does the opposite – it catalyzes the dephosphorylation of $PIP_3$. PTEN is one of the most frequently mutated tumor suppressors that is found in human cancers. So, PI3K and PTEN are parts of a signaling system that can go wrong and lead to cancer" [1].

Not only transcription factors, but also cell surface receptors can be over-expressed. Ligands that bind to the receptors can promote cell growth and division [16]. "For example, the human epidermal growth factor receptor (HER1) appears in many cancer cells. It can be either over-expressed, or it can be expressed in a mutated form that allows it to bind to several ligands that do not bind to the normal HER1. This receptor is a tyrosine kinase which catalyzes its own phosphorylation (auto-phosphorylation). The phosphorylated form binds to two proteins called Grb2 and Sos. These proteins associate with Ras, which activates three proteins on different signaling pathways." [16].

An article in *Scientific American* described advances in treating and curing breast cancer [75]. It described the human epidermal growth factor (hEGF, or HER). HER is an important therapeutic target in many cancers. It is associated with tyrosine kinases, which catalyze the attachment of phosphates to other proteins, which sends growth signals. One of the HER receptors, called HER2 can become locked in the on position in breast cancer cells, causing uncontrolled proliferation. The tyrosine kinase inhibitor called lapatinib is given in combination with Herceptin® to women with HER2-positive cancer. Trastuzumab (Herceptin®), pertuzumab and lapatinib are approved mAbs that act on the HER2/neu receptor [75]. "Lapatinib inhibits the tyrosine kinase activities of the two oncogenes HER2 and EGFR (epidermal growth factor receptor, also known as HER1). Herceptin® is a mAb that binds to the HER2 receptor, reducing cancer cell proliferation. These two drugs act synergistically to inhibit the growth of breast cancer cells. However, cancer cells often develop a resistance to Herceptin®. Tumor cells lose or inactivate the tumor suppressor gene called PTEN (phosphatase and tensin homolog), which normally blocks a PI3K (phosphoinositide 3-kinase) survival pathway, which allows cells to ignore signals that tell them to commit suicide (apoptosis) [75]. PI3K and PTEN are part of a signaling system that can go wrong and lead to cancer [1]. "The third part of this signaling system is a protein kinase called Akt or protein kinase B. Akt indirectly activates another signaling pathway, called mTOR (for mammalian target of rapamycin). The PI3K-PTEN-mTOR pathway is aberrant in many tumors, including breast cancer. So, HER2 expression levels, PTEN activity and the amount of phosphorylated Akt are important biomarkers that help guide treatment of breast cancer. However, when the function of PTEN or mTOR is lost, resistance to herceptin can occur" [1]. So, to prevent breast and other cancers from developing resistance to herceptin, another drug, called Afinitor® has been used [76]. It successfully completed Phase III trials for breast cancer and is now approved for treating renal cancer, non-cancerous kidney tumors caused by the rare genetic disease tuberous sclerosis complex and for breast cancer [76]. It inhibits mTOR.

Another receptor that can be targeted is the estrogen receptor [1]. "Tamoxifen is a drug that binds to estrogen receptors on tumors and decreases DNA synthesis and competes with estrogen for binding to its receptor. Tamoxifen was a standard treatment for breast cancer in patients who have estrogen-sensitive breast cancer. That is, most breast cancers need estrogen to survive, so they can be killed by Tamoxifen, which is an estrogen receptor antagonist in breast tissue. It blocks the binding of estrogen to one of its two types of its receptor (the α receptor)" [1]. In another study, it was shown that the complex formed between tamoxifen and the estrogen receptor represses the transcription of a gene called ERBB2 in patients with elevated levels of ERBB2 in their breast tumors [77].

Other FDA-approved anticancer agents that are hormonal agents include anastrozole, delta-1-testololactone, dimethyltestosterone, dromostanolone propionate, estramustine, ethinyl estradiol, exemestane, fulvestrant, letrozole, megestrol acetate, mitotane, naldrolone, raloxifene, tamoxifen and toremifene [1].

However, Arimidex® (an aromatase inhibitor) has been shown to be superior to either tamoxifen or a combination of tamoxifen and Arimidex® in treating and curing breast cancer. It is given after surgery and radiation therapy [1].

So, the enzyme aromatase (E.C. 1.14.14.1) is a target for therapy in post-menopausal women [1]. "It is in the cytochrome P450 superfamily. It catalyzes the conversion of androgens to estrogens. It catalyzes the conversion of testosterone to estradiol and androstenedione to estrone. That is, it converts the steroid A ring to an phenyl group or benzene ring. About 75% of breast cancers contain elevated levels of estrogens. In these cases, tumor cells that emerge at early stages of breast cancer over-express the gene that codes for aromatase. To treat or possibly even cure breast cancer, drugs (such as Arimidex®) that inhibit aromatase have been used. Letrozole and exemestane are also FDA-approved aromatase inhibitors. They are used as an adjuvant in hormonally responsive breast cancer" [1]. Three more aromatase inhibitors, NeuVax, dHER2 and MVF-HER-2, are in clinical trials [75].

Another target for breast cancer therapy in post-menopausal women is the PI3K/Akt/mTOR survival pathway. As mentioned before, class I PI3K catalyzes the addition of a phosphate to $PIP_2$ to make phosphatidyl inositol 3,4,5-trisphosphate, PI(3,4,5)P3. There are several different isozymes of PI3K. They have been linked to many cellular functions, including cell growth, proliferation, differentiation, motility, survival and intracellular trafficking [78]. "Many of these functions depend on the ability of PI3K to activate protein kinase B, also known as Akt. PI3K activates Akt indirectly. First, PI3K catalyzes the production of PI(3,4,5)P3, which activates Akt. There are three genes in the Akt family, *Akt1*, *Akt2* and *Akt3*. Apoptosis is inhibited by the protein coded by *Akt1*, Akt. Many types of cancer are linked to Akt1. The *Akt1* gene was originally identified as an oncogene. It was found in a retrovirus called Akt8. The Akt protein activates mTOR, or mammalian target of rapamycin. The mTOR protein is a serine/threonine protein kinase. It is the catalytic component of two different signaling complexes, mTORC1 and mTORC2. They integrate signals from multiple inputs, including growth factors, amino acids, and intracellular energy supply, to regulate diverse cellular functions, including transcription, ribosome biogenesis, translation initiation, and autophagic cell death (autophagy). mTOR regulates cell proliferation, motility, and survival. It stimulates protein synthesis and the transcription of DNA into mRNA. It is a negative regulator of autophagy. Compounds that decrease the concentration of inositol or inositol 1,4,5-trisphospate ($IP_3$) can induce autophagy. Tuberous sclerosis 1 (TSC1) and TSC2 are upstream regulators of mTOR that form a functional complex and suppress cell growth by inhibiting mTOR activity. Stimulation by growth factors, such as insulin and the insulin-like growth factor, IGF-1, regulates mTOR signaling through the PI3K-Akt pathway" [78]. Moreover, mTORC1 controls key anabolic and catabolic processes in response to nutrient levels, energy, and growth factors. Often, mTORC1 is deregulated in human diseases, including cancer. So, there are many efforts to develop drugs that inhibit its kinase activity

The mTOR pathway is inhibited by rapamycin, which is made by the bacterium *Streptomyces hygroscopicus* [78]. Rapamycin forms a complex with a small (12-kD) FK506-binding protein (FKBP12), which binds to the catalytic mTOR kinase domain and partially inhibits its activity [1]. Rapamycin has been the subject of much research, not just because of its accepted clinical uses in cancer treatment and organ transplantation, but also for its capacity to prolong the lifespan of yeasts, mice and other organisms [1]. Three drugs that are being evaluated in clinical studies to treat or cure breast cancer are BGT226, BEZ235A and RAD001 [75]. "They inhibit mTORC and PtdIns 3-kinase. Rapamycin itself is also in clinical trials" [75]. However, it had unexpectedly weak anti-cancer activity in some early trials, possibly due to some phosphorylation sites on mTORC targets being resistant to rapamycin treatment. That is, the intrinsic capacity of a phosphorylation site on a target protein to serve as an mTORC1 substrate is a major determinant of its sensitivity to modulators of the pathway [79]. So, mTORC1 effectors can respond differentially to the same signals [79].

Rapamycin itself has poor solubility, so similar compounds, called rapalogs, are being developed [80]. Rapalogs that inhibit the PI3K/Akt/mTOR pathway include sirolimus, temsirolimus and ridaforolimus [80]. Interestingly, rapamycin also significantly increases the lifespan of mice [81]. Inhibition of mTOR may also offer protection against age-related disorders, including cardiovascular disease, cancer and diet-induced obesity [81]. Healthy people should not take rapamycin, though, because it suppresses the immune system.

Another target for breast cancer therapy is the insulin-like growth factor type 1 receptor, or IGF-1 receptor, or IGF-1R [75]. "It is a receptor tyrosine kinase (RTK), with crucial roles in development, aging, and cancer biology. The main signaling pathways mediated by IGF-1 depend on the activation of the phosphatidylinositol 3-kinase (PI3K)-Akt and mitogen-activated protein kinase (MAPK) pathways. High levels of the IGF-1 receptor in blood have been linked with an increased risk of breast cancer. When IGF-1 binds a ligand, it regulates cellular growth and motility, and it protects cells from apoptosis. The IGF-1 receptor protects tumor cells from the effects of chemotherapy and radiation therapy. The drugs called IMC-A12, AMG 479, h7C10 and CP-751,871 are mAbs that target IGF-1 and are in clinical trials" [75]. It should be noted that IGF-1 is also found in milk. Babies and infants need to grow and develop, and IGF-1 helps this. However, children and adults in the USA and many other countries drink much milk. There are many websites on the Internet that claim that drinking milk can cause cancer because of IGF-1. However, IGF-1 is like other proteins. After eating or drinking IGF-1, it is digested and broken down in the stomach. The concentration of IGF-1 in the blood of people who drink milk is much lower than the concentration of IGF-1 in patients with breast cancer.

## Neurodegenerative Diseases

The first year of life is critical in brain development, for the total brain volume doubles, as measured by MRI. This is when the brain is most susceptible to damage by genetic defects and environmental insults. It is also the time in which therapeutic intervention can have its maximum effect. A principal component of the nervous system is the neuron. Neurons are arranged in networks and circuits. The normal human brain has many local regions, or centers, and many pathways between them. The autonomic nervous system is organized into three divisions: the sympathetic, parasympathetic and enteric. These maintenance activities are primarily performed without conscious control or sensation. The sympathetic and parasympathetic nervous systems work to maintain a type of balance. They have opposite effects on the body. The sympathetic division is used in actions requiring quick responses. The parasympathetic division is used in actions that do not require immediate reaction. Messages are sent to and from neurons in the form of primary messengers, called neurotransmitters. L-DOPA is used to treat Parkinson's disease, which affects about 1% of the population over 65. Alzheimer's disease (AD) is the most common neurodegenerative disease.

The human nervous system is a network of molecules, ions, sub-cellular organelles, cells, tissues and organs. Chemical and electrical signals are transmitted from neuron to neuron in complex networks and circuits in the brain. This network and its circular organization are linked with other tissues and organs, including the endocrine system, digestive system, liver and immune system.

The human brain develops slower than the other organs, even though we are born with almost all the neurons that we will ever have. The first year of life is critical in brain development, for the total brain volume doubles, as measured by MRI (magnetic resonance imaging) [82]. This is when the brain is most susceptible to damage by genetic defects and environmental insults. It is also the time in which therapeutic intervention can have its maximum effect [82]. The brain continues to grow during the first two years of life, due to the growth and division of supporting cells (glial cells) in the white matter and there is a large increase in connections (synapses) between neurons in the gray matter. Glial cells are derived from myeloid precursors and migrate into the brain during development. Unlike the most of the rest of the body, it is mostly just these glial cells that are being continuously being broken down and re-made in the brain [83, 84].

Most adult neurons do not undergo mitosis and are not regenerated when they die. Neurogenesis probably does not occur in most regions of the brain, but some data suggest that adult neurogenesis might occur in the neocortex, striatum, amygdala, hypothalamus and brainstem [85]. There is much more evidence that adult neurogenesis occurs in the subgranular zone (SGZ) of the dentate gyrus in the hippocampus (which is involved in learning and memory) and the subventricular zone, or SVZ (in which neural stem cells reside) of the lateral ventricle [86]. Neural stem cells and progenitor cells can respond to external neural activity and differentiate into neurons. This activity-dependent neurogenesis requires $Ca^{2+}$ channels and receptors for the neurotransmitter *N*-methyl-D-aspartate (NMDA). It is called excitation-neurogenesis coupling [87]. That is, neuronal stem cells respond to electrical signals from neighboring neurons by expressing appropriate genes and signaling pathways. This is followed by gamma-amino butyric acid (GABA)-mediated membrane depolarization, which causes an increase in the concentration of intracellular $Ca^{2+}$. The excitation signal activates the transcription factor called NeuroD and promotes neurogenesis. Once the stem cells differentiate into neuroblasts and neurons, excitation by GABA helps the new neurons to be integrated as functional units. Moreover, neural stem cells called type 1 or radial glial-like (RGL) cells can respond to neural activity. RGLs maintain the adult neural stem cell pool in the hippocampus by remaining quiescent. The maintenance and activation of RGLs is controlled dynamically by experience and aging. RGLs that express the protein called nestin can be activated by GABA. The absence of functional $GABA_A$ receptors causes a rapid exit from quiescence and an increased production of RGLs. Interneurons in the stem cell niche that express the $Ca^{2+}$-binding protein parvalbumin are a source of GABA. They are required to maintain RGLs. When GABA signaling is modulated, it affects the generation of more RGLs or causes them to remain quiescent, but not to differentiate [87]. So, the adult brain continues to develop under the influence of electrical activity, while behavior and circuit activity control adult neurogenesis.

Neurogenesis occurs throughout life in the SVZ. Neurons that are generated there migrate to the olfactory bulb, where they become interneurons [86]. Even though most neurons outside the hippocampus, SVZ, and olfacory bulb can't be regenerated, most of the proteins inside them are continuously broken down and re-made [1]. "Also, connections between neurons (synapses) are modified throughout our lives. So, it is striking that people can remember some things for many decades, even though the proteins within the neurons last for a few months or less. Consistent with systems thinking, it is the structure and organization that remains the same, not the molecules. As one thinks about basic building blocks, it is the cell and the organization of networks, and not the individual molecules that build and maintain the nervous system" [1].

Still, an important part of systems thinking (and reductionist thinking) is identifying and characterizing the components of the system [1]. "So, in the science of anatomy, the nervous system is divided into the central and peripheral nervous systems. The brain and spinal cord make up the central nervous system (CNS). The peripheral nervous system (PNS) is made up of all the sensory nerves" [1].

A principal component of the nervous system is the neuron, which is the type of electrically excitable cell that transmits and processes information [1]. "Neurons are arranged in networks and circuits. They communicate with each other and with other types of cells through connections called synapses. Like other human cells, neurons are polarized. The inside of the cell membrane has a slight negative charge (about -70 mV), compared to the outside. When something happens to the membrane, such as electrical stimulation, the voltage difference can change. Non-neuronal cell voltages change rapidly and reach a constant value, where neuronal cells exhibit an action potential, in which the inside becomes positive for a short time and then returns to a negative value" [1].

There are about 10-100 billion neurons in the adult brain, although brain size in humans is not related to intelligence [1]. "This is best illustrated by a disease called hydrocephalus, or water on the brain [1, 88]. Back in the 1940s, when a baby was born with this condition, nothing could be done except let the baby die. Now, doctors can drain the water and babies can live, grow and mature to lead ordinary lives. The disease occurs when the fluid filled spaces in the middle of the brain expand due to increased pressure, pressing the brain against the skull. More than 600 individuals were examined by Dr. Lorber, and about one in ten had more than 90% of the skull volume filled with fluid, and of these, more than half have IQs over 100 [9]. A young man who was a student at Sheffield University had an IQ of 126, earned a first class honors degree in mathematics; and had virtually no brain. The neocortex of the student's brain is a layer about 1/25th of an inch (1 mm) thick, lining his skull. His brain weight was estimated to be about 2/10 of one pound (about 0.090 kg). That is only about 7% of normal brain weight. This student is one of many normal persons with almost no brain" [1, 88]. "Clearly these observational facts show that intelligence and brain size are not related. Normal brain size is not necessary to be of perfectly normal intelligence. What makes man unique is not the size of his brain" [88]. Moreover, young children with autism tend to have larger brains that do neurologically normal children.

In addition, social insects, like honeybees have only $10^6$ neurons, but have complex social behaviors [1]. "Honeybees have a 'dance language' and they can learn abstract concepts, such as 'same' and 'different'. Also, some birds migrate tens of thousands of miles, performing navigational feats that only a few humans can only do with sophisticated technology. Even the 'lowly' insects migrate to their proper locations. Unlike birds, many do not migrate back and forth between two locations in their lifetimes. Instead, many perform inter-generational migration. So, animal behavior and their brains are not like machines, in which 'bigger' means 'better' and 'smaller' means 'inferior'. One might even consider the ongoing battle between pathogenic bacteria and humans. The more antibiotics we use to try to kill them, the more resistant they become, so are we really that much smarter than the bacteria, which have been around for hundreds of millions of years? Modern *Homo sapiens* have been around for a mere 200,000 years and if we don't control our greenhouse emissions and our thermonuclear weapons, we might not survive much longer, even though bacteria will. Still, discussions about the definition of intelligence are best left to the field of psychology. For now, let us focus on human intelligence and look closer at the human brain" [1].

It is the interaction between genetics, environment and epigenetics that make our brains unique [89]. Studies in behavioral genetics indicate that the genetic influence on intelligence appears to be strongest later in life. That is, about 20% of individual differences in intelligence in children can be attributed to genetics, but 80% can be attributed to genetics in some studies of older adults. It is possible that genes nudge people toward choices that shape their environment in a particular way, which in turn affects their intellectual prowess [89].

Next, let us look at Alzheimer's disease (AD), which is the most common neurodegenerative disease. Worldwide, it affects 20 million people and in the USA it affects about five million people, most of them over 70 [14, 90]. A major risk factor is age. The incidence increases from 0.5% per year at age 65 to 8% per year at age 85 years [91]. However, once people reach the age of 95, the incidence of AD actually decreases as they get older [92].

Even in healthy people, some plaques and tangles can appear in the brain [90]. However, they occur much more in AD patients and they have been seen in autopsied brains of people in their 20s [90]. Also, AD is characterized by a progressive decrease in memory, language and other cognitive functions [91]. There is a decrease in behavioral, emotional, interpersonal and social skills. The neuropathological symptoms include an accumulation of extracellular amyloid plaque containing β-amyloid, and intracellular neurofibrillar tangles containing polymerized and hyperphosphorylated tau protein [90, 91]. In AD patients, the abnormal tau protein accumulates, and tangles the microtubules that are essential for neurons to communicate with each other and to receive their needed nutrients and chemical messengers [92].

It is likely that AD is actually several diseases with several possible genetic causes [91, 92]. Still, people who are homozygous (have two copies) of the ApoE epsilon-4 allele (APOE-4) have an increased risk of AD. There are three APOE genes, called APOE-2, 3 and 4 [92]. APOE-4 is linked to AD. People with one copy of APOE-4 have three times the risk of getting AD as people with only APOE-2 and 3. People who have two copies of APOE-4 (homozygous) have eight times the risk of getting AD [90]. ApoE is a protein that influences a person's ability to develop AD [90]. It codes for an abnormal version of a protein called β-amyloid precursor protein, or APP [40]. APP catalyzes the hydrolysis of a large protein that is a precursor of β-amyloid. In healthy people, APP is a transmembrane glycoprotein with an unknown, but apparently essential function. Healthy β-amyloid protein is water-soluble, but improperly cut β-amyloid protein is not.

Also, about 5-10% of AD patients get AD before they turn 65. This is called early onset AD. It can be caused by a defective gene on chromosome 21, which exists as three copies (trisomy) in Down's syndrome patients. Such patients develop abnormal β-amyloid plaques. AD patients don't just make more APP, but they cut (hydrolyze) the precursor protein wrong [92]. When APP is cut wrong, it becomes sticky and clumps together, forming plaques that are neurotoxic. The plaques stimulate the immune system in the brain to mount an inflammatory response, further damaging the neurons. There are at least three genes that contribute to plaque formation, APP, presenilin 1 and presenilin 2. Also, latent herpes virus may facilitate the development of AD in people who have the defective ApoE gene [91].

However, the behavioral symptoms of Alzheimer's disease can be prevented in some cases by staying socially connected, keeping the heart (cardiovascular system) in good shape, avoiding strokes, obtaining a higher education and avoiding depression [92]. "There was an epidemiological study of retired nuns who have agreed to take intelligence tests every year while they are alive and donate their brains to AD research after they died. Strokes and untreated clinical depression emerged during the study as important risk factors for AD. During autopsies, some of the nuns' brains exhibited the amyloid plaque and neurofibrillar tangles in the entorhinal cortex and hippocampus typical of AD, even though they don't show any symptoms of it when they are alive. Doctors and scientists use a scale called the Braak scale to classify different levels of AD. If there are no neurofibrillar tangles and no plaques, there is no AD. If there are tangles in the entorhinal cortex, it is stage I or II of AD. The entorhinal cortex is located at the base of the skull and is important for memory. If there are tangles higher and deeper in the brain, and they reach the hippocampus and neighboring regions, AD has reached stages III or IV. The hippocampus is critical to learning and memory. It processes and stores new information. If the tangles and plaques reach the neocortex, the AD is at stages V or VI. Still, there was one nun who stayed very connected to her sister nuns, had never suffered a stroke or depression and had a higher college education. Upon her death, an autopsy showed that she had an unusually small brain and had many tangles in her entorhinal complex and hippocampus, but not in her neocortex. She showed no symptoms of AD when she was alive" [92].

On the other hand, it is quite common for widows to develop AD after their husbands die and they are overcome with loneliness. Still, even though there is much a person can do to prevent the outward symptoms of AD, much research continues on the biochemical and genetic components.

Also, the brains of nuns with AD had less folic acid and less lycopene in them than nuns who died with no symptoms of AD [92]. This does not necessarily mean that folic acid and lycopene prevent AD. It is possible that oxidative damage consumes the folic acid and lycopene in AD patients. In any case, defective amyloid proteins are produced [92].

Amyloid-β is 40 – 42 amino acids long and it is produced by the hydrolysis of the much larger amyloid precursor protein, or APP, which is in the cell membrane. Amyloid-β is a secreted protein and it has only a partially understood function in healthy brains [93]. It binds $Cu^{2+}$ and other transition metals. When $Cu^{2+}$ is bound, it gives amyloid-β its antioxidant properties [93]. The N-terminus of APP is located in the extracellular domain and its C-terminus is inside the membrane. Two different enzymes catalyze the hydrolysis of the N- and C-termini. They are called β-secretase and γ-secretase, respectively. A third enzyme, called α-secretase catalyzes the hydrolysis of APP between amino acid residues 16 and 17, which prevents the formation of amyloid-β. The gene that codes for the APP protein (called the APP gene) has been mapped to chromosome 21. Three copies of this chromosome exist in people with Down's syndrome, who also have amyloid plaques in their brains. A small percentage of people who have AD (the Volga Deutsch) have mutations in some of their genes that might help cause the disease. However, mutations of the APP gene cause only a minority of these familial AD cases. Linkage studies found a major disease locus on chromosome 14 and positional cloning led to the identification of mutations in the *presenilin-1* gene. Mutations in the *presenilin-1* gene are the most common cause of familial AD [94].

The presenilin protein is a part of an aspartyl protease complex that works with γ-secretase to catalyze the hydrolytic cleavage of APP [94]. Like other aspartyl proteases, it contains an aspartyl (aspartate) residue in its active site. In preclinical cases with *presenilin-1* mutation, deposits of amyloid-β 42 are seen. Moreover, studies with transgenic animals indicated that *presenilin-1* mutations can also lead to hyperphosphorylated tau protein even when there are no amyloid-β plaques [94]. Recent evidence indicated that the presenilin proteins and the normal amyloid-β may play important roles in maintaining the proper amount of intracellular $Ca^{2+}$ in neurons [94]. Disease-causing mutations may form channels in the cell membrane, thus disrupting the proper regulation of intracellular $Ca^{2+}$ [94]. Another protein (tau) is also affected.

The tau protein is involved in the assembly and stabilization of microtubules [90]. "In the human brain, six isoforms of tau are produced by splicing different portions of mRNA. Filamentous tau deposits are also formed in some other neurodegenerative diseases, which do not have amyloid-β plaques. Tau pathology appears first in the part of the brain called the transentorhinal region. From there it spreads to the hippocampus, amygdala and finally the cerebral cortex. Amyloid-β appears in the neocortex. Both types of inclusion happen independently, with tangles appearing first. In fact, neurodegeneration in AD starts to occur 20 to 30 years before the appearance of the first clinical symptoms. The first phase is called amnesiac mild cognitive impairment (aMCI). Its neuro-pathological features are intermediate between normal aging and AD. Tau deposits are abundant in the entorhinal cortex and hippocampus and some amyloid-β plaques are in the neocortex. There is a chemical called Pittsburgh compound B (PIB), which has been used to visualize amyloid-β plaques in both AD patients and normal elderly people. This may help to identify people at risk for AD and help in efforts to develop ways to prevent it" [90].

Also, one of the pathways of neuronal death in AD is mediated by free radical damage and reactive oxygen substances (ROS), produced by NADPH oxidase and derivatives of arachidonic acid [88, 91]. It is possible that amyloid-β 42 induces ROS production and stimulates the release of arachidonic acid through NMDA receptors [88]. As a result, amyloid-β may mediate oxidative damage in AD by acting on the NMDA subtype of ionotropic glutamate receptors. The brain is especially vulnerable to oxidative damage because it uses so much oxygen in its metabolism, and it has a relatively large amount of polyunsaturated fats that are easily damaged by ROS. In addition to aging, Down's syndrome, vascular disease, smoking tobacco and diabetes are important risk factors for AD and all of these involve oxidative damage. Antioxidants, such as NSAIDs, curcumin, apocynin, docosahexaenoic acid (DHA) and other omega-3 fats may help prevent AD [95-99]. However, some studies indicated that NSAIDs do not prevent AD [100].

More recently, it has been shown that the root cause of AD may be oxidative stress [101], which occurs before the buildup of amyloid plaques. There is a long dormant period of gradual oxidative damage (smoldering inflammation) that precedes and eventually leads to the seemingly sudden appearance of clinical and pathological symptoms of AD symptoms, including amyloid-β (Aβ) deposition, neurofibrillary tangle formation, metabolic dysfunction, and cognitive decline [102]. In fact, Aβ has been shown to increase the amount of iron and oxidative stress [103]. This might be a common factor in not just AD, but also Parkinson's and Huntington's disease [104]. It is also possible that the soluble form of Aβ (amino acids

25-35) may cause neurotoxicity in the hippocampus before the appearance of amyloid plaques [105].

## REFERENCES


[1] Smith, R. E. *Medicinal Chemistry – Fusion of Traditional and Western Medicine,* Bentham Science, Sharjah, U.A.E., 2013.
[2] Farese Jr, R. V., Cases, S., J. Smith, S. J. *Current Opin. Lipidol.* 2000, 11, 229-234.
[3] Lazar, M. A. *Sci.* 2008, 331, 1048-1049.
[4] Maratos-Flier, E. *Nature Med.* 2008, 14, 604-606.
[5] Sun, K. et al. *Proc. Natl. Acad. Sci.* 2012, 109, 5874-5879.
[6] Mayuoni-Kirshinbaum, L., Daus, A., Porat, R. *Food Sci. Technol.* 2013, 48, 1569-1578.
[7] P Melgarejo, P., Artés, F. *J. Sci. Food Agric.* 2000, 80, 1452-1454.
[8] Lua, J. Wei, Y., Yuan, Q. *J. Chromatogr. B* 2007, 857, 175-179.
[9] Sack, C. et al. *J. Agric. Food Chem.* 2013,
[10] Dawkins, R. *The Selfish Gene*, Oxford Press, Oxford, pp. 39, 214, 215 1976.
[11] Noble, D. *The Music of Life: Biology Beyond the Genome*, Oxford University Press, Oxford 2006.
[12] Vaidya, N et al. *Nature* 2012, 491, 72-77.
[13] Lai, F. *Nature* 2013, 494, 497-501.
[14] Quinn, A. E. et al. *Sci.* 2007, 316, 411- 2007.
[15] McClintock, B. *Proc. Natl. Acad. Sci.* 1950, 36, 344-355.
[16] Corey, E. J., Czako B, Kürti, L. *Molecules and Medicine*, John Wiley & Sons, New York 2007.
[17] Adler, E. M. *Sci. Signal.* 2010, 3, 103 eg1.
[18] SA Biosciences, Pathways online, http://www.sabiosciences.com/pathwaysonline/ 2012.
[19] Kholodenko, B. N., Hancock, J. F., Kolch, W. *Nature Rev. Mol. Cell. Biol.* 2010, 11, 414-426.
[20] Pfleger, C. M. *Sci. Signal.* 2011, 4 (163), pe12.
[21] Lorentzen, A. et al. *Sci. Signal.* 2010, 3, ra68.
[22] Fang, J. Y., Richardson, B. C. *Lancet Oncol.* 2005, 6, 322.
[23] Davis, R. J. *Sci.* 2012, 337, 1178-1179.
[24] Turjanski, A. G., Vaqué, J. P, Gutkind, J. S. *Oncogene* 2007, 26, 3240.
[25] Vermeij, R. et al. *J. Biomed. Biotechnol.* 2011, 2011, 1-11.
[26] Kruse, J.-P., Gu, W. Cell 2009; 137: 609-622.
[27] Istvan, E. S., Deisenhofer, J. Sci. 2001, 292, 1160-1164.
[28] Schulz, M. M. P. et al. *Proc. Natl. Acad. Sci.* 2012, 109, E2665-2674.
[29] Jarvis, L. M. *Chem. Eng. News* 2008, 86 (May19), 29.
[30] Dodd-o, J. M. et al. *AJP-Heart* 2004, 287, 927-936.
[31] Mantovani, A. et al. *Nature* 2008, 454, 436-444.
[32] Sullivan, J. L. *Lancet* 1981, 1, 1293-1294.
[33] Kell, D. B. *Arch. Toxicol.* 2010, 84, 825–889.

[34] Medzhitov, R. *Nature* 2008, 454, 428-435.
[35] Libby, P. *Amer. J. Clin. Nutr.* 2006, 83 (supplement), 456S-460S.
[36] Barnig, C. et al. *Sci. Transl. Med.* 2013, 5 (174ra26), 1-11.
[37] van Greevenbroek, W., Schalkwijk, C. G., Stehouwer, C. D. A. *Neth. J. Med.* 2013, 71, 174-187.
[38] Gorman, R. M. *Proto Dispatches Front. Med.* 2008, 26-31.
[39] Crunkhorn, S. *Nature Rev. Drug Disc.* 2013, 12, 261.
[40] Schlosser, E. *Fast Food Nation*, Houghton-Mifflin, New York, 2001.
[41] Tan, Z. S. et al. *Neurol.* 2007, 68, 1902-1908.
[42] McGeer, E. G, McGeer, P. L. *Mol. Interventions* 2001, 1, 22-29.
[43] Wang, Y. et al. *Exp. Neurol.* 2005, 193, 75-84.
[44] Waris, G., Ahsan, H. J. *Carcinogen.* 2006; 5, 14.
[45] Belkaid, Y., Grainger, J. *Sci.* 2013, 342, 432-433.
[46] Shan, M. et al. *Sci.* 2013, 342, 447-453.
[47] Geismann, F. et al. *Sci.* 2010, 327, 656.
[48] Rosengren, A. H. et al. *Sci.* 2010, 327, 217-220.
[49] Ptinopoulou, A. G., Pikilidou, M. I., Lasaridis, A. N. *Hypertension Res.* 2013, 36, 91–101.
[50] Yusef, S. et al. *Lancet* 2009, 373, 1341-1351.
[51] Paneni, F., Beckman, J. A., Creager, M. A., Cosentino, F. *Eur. Heart J.* 2013, 34, 2436-2443.
[52] Capra F. *The Web of Life*, Doubleday, New York, 1996.
[53] Lyssiotis, C. A., Cantley, L. C. *Cell* 2012, 151, 1155-1156.
[54] Meacham, C. E., Morrison, S. J. *Nature* 2013, 501, 328-337.
[55] Bedard, P. L., Hansen, A. R., Ratain, M. J., Siu, L. L. *Nature* 501, 355-364.
[56] Burrell, *R. A.*, McGranahan, N., Bartek, J., Swanton, C. *Nature* 2013, 501, 338-345.
[57] Klein, C. A. *Nature* 2013, 501, 365-372.
[58] Plaks, V., Koopman, C. D., Werb, Z. *Sci.* 2013, 341, 1186-1188.
[59] Marx, J. *Sci.* 2008, 320, 38-41.
[60] Folkman, J. *Nature* 2007, 6, 273-286.
[61] Junttila, M. R., Frederic, J. de Sauvage, *F. J. Nature* 2013, 501, 346-354.
[62] Zindl C. L., Chaplin, D. D. *Sci.* 2010, 328, 697.
[63] Tlisty, T. *Nature* 2008, 453, 604-605.
[64] McMillin, D. W., Negri, J. M., Mitsiades, C. S. *Nat. Rev. Drug Disc.* 2013, 12, 217-228.
[65] Zhang, Y. et al. Cancer Res 2012; 72: 5198–5208.
[66] Baselga, J. *Sci.* 2006, 312, 1175-1178.
[67] Adler, E. M. *Sci. Signal.* 2010, 3, 103 eg1.
[68] Vermeij, R. et al. *J. Biomed. Biotechnol.* 2011, 2011, 1-11.
[69] Kruse, J.-P., Gu, W. *Cell* 2009, 137, 609-622.
[70] Gardner, L., Lee, L., Dang, C. The c-Myc Oncogenic Transcription Factor in *Myc Oncogene, Encyclopedia of Cancer*, $2^{nd}$ ed. Bertino, J. R. ed. Elsevier, New York, 2002.
[71] Vita, M., Henriksson, M. *Seminars Cancer Biol.* 2006, 16, 318–3.
[72] Dang, C. V. *Cell* 2012, 149, 22-35.
[73] Burrell, R. A. et al. *Nature* 2013, 494, 492-498.

[74] Janssen, A., Medema, R. H. *Nature* 2013, 494, 439-441.
[75] Esteva, F. J., Hortobagyi, G. N. *Sci. Amer.* 2008, 298, 61-65.
[76] Afinitor Official website: http://www.afinitor.com/index.jsp?site=PC018060&source=01030&irmasrc=ONCWB0042
[77] Hurtado, A. et al. *Nature* 2008, 456, 663-667.
[78] Lee, C.-H., Inoki, K., Guan, K.-L. *Ann. Rev. Pharmacol. Toxicol.* 2007, 47, 443-467.
[79] Kang, S. A. et al. *Sci.* 2013, 341, 6144.
[80] Blay, J.-Y. *Ann. Oncol.* 2011, 22, 280-287.
[81] Kaerberlein, M., Kennedy, B. K. *Nature* 2009, 460, 361.
[82] Knickmeyer, R. C. et al. *J. Neurosci.* 2008, 19, 12176-12182.
[83] Lie, D. C. et. al. *Ann. Rev. Pharmacol. Toxicol.* 2004, 44, 399-421.
[84] Bhardwaj, R. D. et. al. *Proc. Natl. Acad. Sci.* 2006, 103, 12564-12568.
[85] Glasper, E. R., Leuner, B., Gould, E. *Nature Neurosci.* 2008, 11, 731.
[86] Ming, G., Song, H. *Ann. Rev. Neurosci.* 2005, 28, 223-250.
[87] Hsieh, J., Schneider, J. W. *Sci.* 2013, 339, 1534-1535.
[88] Lewin, R. *Sci.* 1980, 210, 1232-1234.
[89] Van Essen, D. *Sci.* 2012, 338, 35-36.
[90] Goedert, M., Spillatini, M. G. *Sci.* 2006, 314, 777-781.
[91] Praticó, D., Delanty, N. *Physiol. Med.* 2000, 109, 577-585.
[92] Snowden, D. *Aging with Grace: What the Nun Study Teaches Us about Leading Longer, Healthier and More Meaningful Lives*, Bantam Books, New York, 2001.
[93] Atwood, C. S. et al. *Brain Res. Rev.* 2003, 43, 1-16.
[94] Marx, J. *Sci.* 2007, 318, 384-385.
[95] Yang, F. et al. *J. Biol. Chem.* 2005, 280, 5892-5901.
[96] Lim, G. P. J. Neurosci 2005; 25: 3032-3040.
[97] Hashimoto, Y. *J. Neurochem.* 2003, 84, 864-877.
[98] Vlad, S. C. et al. *Neurol.* 2008, 70, 1672-1677.
[99] Szekely, C. A. *Neurol.* 2008, 70, 17-24.
[100] Arvanitakis, Z. et al. *Neurol.* 2008, 70, 2219-2225.
[101] Bonda, DJ. et al. *Neuropharmacol.* 2010, 59, 290-294.
[102] Smith, M. A. et al. J. Alzheimers Dis. 2010, 19, 363–372
[103] Wan, L. et al. *Free Rad. Biol. Med.* 2011, 50, 122–129.
[104] Peña, F. et al. *Hippocamp.* 2010, 20, 78-96.
[105] Kurnellas, M. P. et al. *Sci. Transl. Med.* 2013, 5(179ra42), 1-11.

# AUTHOR CONTACT INFORMATION

**Dr. Robert E. Smith, Ph.D.**
Assistant Professor
Park University and Science
1959 W 139th Terr
Leawood, KS 66224
Tel: 913-814-7663
E-mail: rmaracuja@gmail.com

# INDEX

## A



## B



## C



## D



## E



## F



## G



## H

## I



## J



## L



## M



## N



## O



## P



## R



## S



## T



## U